BIOTECHNOLOGY
AND THE FUTURE OF SOCIETY
CHALLENGES AND OPPORTUNITIES

BIOTECHNOLOGY AND THE FUTURE OF SOCIETY

CHALLENGES AND OPPORTUNITIES

THE EMIRATES CENTER FOR STRATEGIC STUDIES AND RESEARCH

THE EMIRATES CENTER FOR STRATEGIC STUDIES AND RESEARCH

The Emirates Center for Strategic Studies and Research (ECSSR) is an independent research institution dedicated to the promotion of professional studies and educational excellence in the UAE, the Gulf and the Arab world. Since its establishment in Abu Dhabi in 1994, ECSSR has served as a focal point for scholarship on political, economic and social matters. Indeed, ECSSR is at the forefront of analysis and commentary on Arab affairs.

The Center seeks to provide a forum for the scholarly exchange of ideas by hosting conferences and symposia, organizing workshops, sponsoring a lecture series, and publishing original and translated books and research papers. ECSSR also has an active fellowship and grant program for the writing of scholarly books and for the translation into Arabic of work relevant to the Center's mission. Moreover, ECSSR has a large library including rare and specialized holdings and a state-of-the-art technology center, which has developed an award-winning website that is a unique and comprehensive source of information on the Gulf.

Through these and other activities, ECSSR aspires to engage in mutually beneficial professional endeavors with comparable institutions worldwide, and to contribute to the general educational and scientific development of the UAE.

First published in 2004 by
The Emirates Center for Strategic Studies and Research
PO Box 4567, Abu Dhabi, United Arab Emirates

E-mail: pubdis@ecssr.ac.ae
pubdis@ecssr.com

Website: http://www.ecssr.ac.ae
http://www.ecssr.com

Distributed by The Emirates Center for Strategic Studies and Research

British Library Cataloguing in Publication Data
A catalogue record for this book is available from the British Library

Library of Congress Cataloguing in Publication Data
A catalogue record for this book is available from the Library of Congress

ISBN 9948-00-509-0 hardback edition
ISBN 9948-00-508-2 paperback edition

Printed by Ithaca Press, 8, Southern Court, South Street,
Reading RG1 4QS, United Kingdom

Contents

FIGURES AND TABLES

FIGURES

TABLES

Abbreviations and Acronyms

AAAS	American Association for the Advancement of Science
ABM	anti-ballistic missile
AC	hydrocyanic acid
ADHD	attention deficit/hyperactivity disorder
AIDS	acquired immune deficiency syndrome
ART	assisted reproductive technology
AML	acute myeloid leukemia
ASRM	American Society for Reproductive Medicine
BMS	Bristol-Myers Squibb
BWC	Biological Weapons Convention
CAT	computerized axial tomography
CD	compact disk
CDC	Centers for Disease Control
CEO	Chief Executive Officer
CEDD	Centers of Excellence for Drug Discovery
CGIAR	Consultative Group on International Agricultural Research
CHI	Cambridge Healthtech Institute
CK	cyanogen chloride
CML	chronic myeloid leukemia
CRO	contract research organization
CSO	contract sales organization
CT	computed tomography
CTBT	Comprehensive Test Ban Treaty
CWC	Chemical Weapons Convention
CX	phosgene oxime
DOE	Department of Energy
DNA	deoxyribonucleic acid
EC	embryonal carcinoma (cells)
EG	embryonic germ (cells)
ES	embryonic stem (cells)
ECG	electrocardiogram
EPA	Environmental Protection Agency
EU	European Union
FAA	Federal Aviation Administration

FACS	fluorescence-activated cell sorting
FAO	Food and Agriculture Organization
FDA	Food and Drug Administration
FDAMA	Food and Drug Administration Modernization Act
FISH	fluorescent in situ hybridization
GCC	Gulf Co-operation Council
GDP	gross domestic product
GMO	genetically modified organism
GPS	global positioning system
GSK	GlaxoSmithKline
GTP	guanosine triphosphate
hES	human embryonic stem cells
HIPAA	Health Insurance Portability and Accountability Act
HIV	human immunodeficiency virus
HTS	high-throughput screening
ICBM	intercontinental ballistic missile
ICM	inner cell mass
ICS	Institute for Civil Society
IED	improvised explosive devices
IND	investigational new drug
IP	intellectual property
IQ	intelligence quotient
ISAAA	International Service for Acquisition of Agribiotech Applications
IT	information technology
IVF	in vitro fertilization
JNJ	Johnson & Johnson
M&A	merger and acquisition
MAb	monoclonal antibody
MAP	mitogen-activated protein
MEM	micro-electrical machine
MIT	Massachusetts Institute of Technology
MRI	magnetic resonance imaging
MTA	Managing the Atom (project)
NAS	National Academy of Sciences
NBAC	National Bioethics Advisory Commission
NCE	new chemical entity

NDA	new drug application
NIBR	Novartis Institute of Biomedical Research
NIH	National Institutes of Health
NME	new molecular entity
NPT	Non-Proliferation Treaty
NRC	National Research Council
NREL	National Renewable Energy Laboratory
NTI	Nuclear Threat Initiative
OECD	Organization for Economic Cooperation and Development
PAPSAC	Private and Public, Scientific and Consumer Policy Group
PFIB	perfluoroisobutylene
PC	personal computer
PDR	Physicians' Desk Reference
PET	positron emission tomography
PGCs	primordial germ cells
PGD	pre-implantation genetic diagnosis
PHA	polyhydroxyalkanoate
PLA	polylactides
PPP	purchasing power parity
PTO	Patent and Trademark Office
R&D	research and development
ROM	read-only memory
SALT	Strategic Arms Limitation Talks
SCIDS	severe combined immunodeficiency syndrome
SCNT	somatic cell nuclear transfer
SNP	single nucleotide polymorphism
TBC	Texas Biotechnology Corporation
UAE	United Arab Emirates
UCLA	University of California at Los Angeles
UNESCO	United Nations Educational Scientific and Cultural Organization
USAMRICD	US Army Medical Research Institute of Chemical Defense
USDA	United States Department of Agriculture
WHO	World Health Organization
WIPO	World Intellectual Property Organization
WMD	weapons of mass destruction

Foreword

The ongoing biotechnological revolution has the potential to impact dramatically on our lives. New techniques and processes being developed in medicine, healthcare, agriculture and security will undeniably transform these fields. Possible future scenarios include the rectifying of genetic disorders prior to birth, the growing of human tissues and organs, and even extending the human lifespan. The so-called 'Green Revolution,' which relies increasingly on genetic modification, may achieve increased food production on less land at lower costs, thus serving to alleviate world hunger. Genetic fingerprinting and sophisticated identification and detection techniques will greatly facilitate law enforcement and enhance security services.

Despite all these promising prospects, the supposed 'inevitability' of the biotechnology revolution and its manifold effects need to be subjected to scrutiny. It is unacceptable to allow the biotechnology revolution to merely 'happen,' without understanding its broader impact. Biotechnological methods employed in the growing of new kinds of plants, tissues, organs and organisms need careful evaluation. Are they really beneficial advances that enhance the quality of life while safeguarding human dignity and ensuring the future of this planet? Or are they merely news-making scientific discoveries that may entail hidden and uncontrollable risks? Will manipulation of the genetic code of organisms cause unpredictable disruptions to our ecosystem, undermine living standards and threaten our very survival as a species?

The field of biotechnology has the undeniable capacity to affect several aspects of life. However, the degree of its impact will depend to a great extent on individual and social choices, which in turn must be based on objective evaluation. To facilitate such an assessment, the Eighth Annual Conference of the Emirates Center for Strategic Studies and Research, entitled *Biotechnology and the Future of Society: Challenges and Opportunities* assembled a panel of internationally renowned experts in Abu Dhabi, UAE, from January 11–13, 2003. This forum examined the latest advances in the life sciences and discussed their social, ethical, legal, industrial and health implications.

The informative conference presentations compiled in this volume thus highlight the major opportunities and challenges stemming from the biotechnological revolution. It is hoped that this book will help to impart the necessary understanding about biotechnology and its wider effects to enable responsible decision making in this significant and rapidly developing field. The Center would like to record its appreciation to the eminent speakers who participated in the conference and shared their valuable expertise. A word of thanks is also due to ECSSR editor Mary Abraham for coordinating the publication of this book.

Jamal S. Al-Suwaidi, Ph.D.
Director General
ECSSR

Section 1

INTRODUCTION

Biotechnology and the Future of Society: An Overview

Jamal S. Al-Suwaidi

The world is about to exit the Information Age and enter the new era of "bioterials." The marvels of the Bioterials Age will be more global in their impact than the Internet. Its products will be more important than fire, the wheel or the car, and faster and more productive than today's biggest supercomputers. The bioterials era will generate more new knowledge in a shorter period than history's collective wisdom, and the power of its technologies will eclipse that of the combined armies of the world.[1]

Whether one agrees entirely or not with the sensational claims above made by Richard W. Oliver, the general consensus is that biotechnology is here to stay. No individual or government in the history of technological innovation has been able to halt its progress. The nineteenth century witnessed the age of chemistry, and the twentieth placed physics at the center of innovation. The twenty-first century will see biology at the core of scientific and technological endeavor. The nature of the new "Age of Biology," an umbrella term for the various facets of the new biology-based scientific and technological era, will ensure its survival and expansion.

The certainty of its existence, the hopes and anxieties around its pervasive influence, and the need to determine its direction for the benefit of humanity compels every individual to understand the essence and complexity of the biotechnology revolution. It is necessary that the general public should not remain passive spectators to the impact of this new technology, assuming the inevitability of all its manifestations, but should educate themselves about its nature and possible influence. Furthermore, it is imperative that people the world over should make their voices heard on the issues of the new age, so that they may be beneficiaries of its technologies, rather than passive recipients of their impact.

Jeremy Rifkin, in his book *The Biotech Century*, observes that we live the legacy of unexamined issues of two earlier scientific revolutions in physics and chemistry in the nineteenth and twentieth centuries. "Both brought great benefits to humankind along with equally significant problems."[2] He goes on to explain that these problems are an "environmental, social and economic bill that is also the shared legacy of the modern age."[3]

It is to avoid falling into the same trap of facing the consequences of an unexamined response, that The Emirates Center for Strategic Studies and Research hosted its Eighth Annual Conference in Abu Dhabi, United Arab Emirates, entitled *Biotechnology and the Future of Society: Challenges and Opportunities*. Some of the foremost experts in the field were invited to share their knowledge and perspectives, and this book represents their collective insight. The volume is intended to present the reader with an opportunity to examine the key developments within the broad spectrum of the new "Age of Biology" and to construct an informed opinion on issues that are certain to touch the life of every individual.

Thus far, media and public attention has focused on the genetic engineering aspects of the biotechnology revolution. Debate has been lacking on other potentially beneficial areas of the developments in the life sciences, such as the more integrative and embedded technological applications, those that are more sensitive to ecosystem dynamics and interrelationships.[4] Nevertheless, the emphasis on genetic engineering aspects of the biotech revolution may well be justified. As John Gearhart mentions in his chapter "Stem Cell and Cloning Research: Implications for the Future of Humankind," stem cell research is arguably the most important subject in biology today, because this issue has sharply focused society's attention on the ethical, political and economic aspects of this cutting-edge research.

In their book *2020 Vision: Transform Your Business Today to Succeed in Tomorrow's Economy*, Stan Davis and Bill Davidson state that the science of genetics is providing the predictive bridge between biology and technology.[5] They maintain that science has only been able to usher in new technology when it has been able to become predictive. Until breakthroughs occurred in understanding DNA and carrying out genetic engineering, biology had failed to provide that predictive bridge that would enable technological innovation and the dawn of a new economic age.

The biotechnology revolution, in simple terms, represents a method of manipulating the elements of life.[6] It began in 1973 with Watson and Crick's success in splicing the DNA of two distinct organisms, referred to as "recombinant technology." This technology is at the heart of the biotech revolution. It will usher in the Biotech Age and absorb a substantial part of the global intellectual capital in the first century of the new millennium. One of the principal factors that will ensure both its global expansion and its intimate individual reach is the fact that the key element of the previous era, the Information Age, has now merged with the powerful innovation of the Biotech Age – computers have joined forces with biology, ushering in the era of bioinformatics. As Layne, Beugelsdijk and Patel state in their book *Firepower in the Lab*, "Scientists have fast computers, database technologies and Internet connections (i.e. powerful "digital" firepower) for analyzing enormous problems in medicine and biology."[7] Furthermore, as Gearhart explains, the biosciences are based on teamwork and worldwide collaboration. Clearly, information technology has played and will continue to play a substantial role in the advances of the Biotech Age.

In his book *The Biotech Century*, Jeremy Rifkin states that, "After 40 years of running on parallel tracks, the information and life sciences are fusing into a single powerful technological and economic force."[8] This merger of science and technology is already allowing for the processing of vast amounts of biological and biotech data. For example, as Michio Kaku mentions in his chapter "Advances in the Field of Biotechnology," initial fears that the Human Genome Project would be too expensive and time-consuming have proved incorrect. Gene sequencing has been automated, with computer power doubling every 18 months and results being shared instantly over the Internet.

The rapid dissemination of knowledge is not limited to the scientific community. Computer power has also made this information accessible to the public. This acquisition of knowledge will inevitably result in the empowerment of the individual, and give rise to social change. Such transformation will, in the words of Thomas C. Wiegele, in his book *Biotechnology and International Relations*, "create new political realities that will have to be addressed by national governments as well as by the international community."[9]

Among such changes is the already discernible influence on the world economic system. Biotechnology patent approvals have shown an incremental increase of almost sevenfold in the last twenty years, and knowledge of biotech materials is forecasted to double daily early in this century – both being indicators of the economic potential of the biotech age.[10] In his chapter on "Investing in the Biotechnology Industry: The Role of Research and Development," Andrew Greene makes the statement that several factors indicate a huge market for biotechnology innovations in the sphere of health. Such factors include an aging global population, increasing expendable income for healthcare, the fact that antibiotic-resistant infectious diseases are making a comeback and no cure has yet been found for the major cancers.

Gregory Stock, in his chapter "Biotechnology and the Future of Medicine," states that the real tests of how biotechnology will reshape medical practice will come from developments that threaten the fundamental structure of medicine by blurring the lines between therapy and enhancement, prevention and treatment, need and desire. He adds that one of the most significant changes ahead will be a massive shift towards preventive medicine. Glenn McGee provides further indications of the health sector potential in his chapter "Ethical, Legal and Social Issues in the Field of Biotechnology." McGee states that in the United States 33 percent of all visits to Internet sites are for healthcare information, while 37 percent of such visits in the Arab world are for the same purpose.

In general, biotech innovations hold out the promise of great profit. This is because the impact of the new technology on many fields of human endeavor and concern – including health, agriculture, defense, commerce and industry – is both global in scale and pervasive in depth.

These new technologies also have a strong spillover potential into many of the more traditional industries, what could be termed as "enabling technologies." It is estimated that, in less than one generation, almost every company worldwide will either be involved in the development and/or use of the new biotechnologies or depend on them for its survival.[11] For over three decades, venture capitalists have invested substantially in biotechnological innovation. The rewards are potentially enormous. Such prospects for financial and material benefits indicate that multinational corporations are likely to be key players in the new age, driving social change and compelling political responses.

For these responses to be informed and socially rooted, it is crucial that knowledge of the various aspects of the life sciences be promoted at ground level. The new term "bioliteracy" has already been coined, to replace "information literacy" as the essential form of literacy required for an educated opinion on the Biotech Age.[12] In their chapter "Will Life Sciences be a Driving Force of the Twenty-First Century Economy?" Juan Enriquez-Cabot and Helen Quigley state that societies that do not become science-literate will become poor very quickly.

Lack of public awareness regarding the true ambit of new developments has led to the understanding of biotechnology being hijacked by one of its sub-fields, genetic engineering, notably with the advent of Dolly, the cloned sheep, and then later by disconcerting claims of human cloning. This narrow perception of the new Biotech Age does it a disservice, spreading fears of a world of cloned humans, animals and plants, while obscuring its many other clearly beneficial facets. For example, genetic manipulation holds strong potential for the treatment of diseases in humans and, as Michio Kaku mentions in his chapter, gene therapy may well become a key weapon against disease.

Another area of human endeavor that has already benefited from genetic manipulation has been agriculture. For centuries, humans have contrived to mix strains of crops to produce the most resilient and productive strains. However, it has taken many generations of crop cycles to improve on the plant species. What biotechnology can now do is to produce the same result by conflating generations of experiments into a single exercise in the laboratory, which results in a "simultaneous increase in quality, quantity and geographic scope of production."[13]

In his chapter "Biotechnology and the Agricultural Industry of the Future," Ray Goldberg makes the statement that the so-called "Green

Revolution" to reduce world hunger relies to a large extent on genetic modifications that enable crops to resist extreme climatic and soil conditions. In these times of genetic revolution, issues of the global food system are not just those relating to productivity but also to the environment, world trade, nutrition, food safety, health and the political stability of nations and regions. They include issues of access to technology, and intellectual property safeguards. If common ground cannot be found on these critical issues, economic development in both the developing and the developed world will be threatened, with the burden falling on the poorer nations and populations. In their book, *Biological Resource Management: Connecting Science and Policy*, Balázs et al. maintain:

> Although plant transformation was initially experimental, the potential commercialization of new improved varieties was early realized and a fast-growing international ag-biotech [agricultural-biotech] market has already been formed.[14]

The same advances that biotechnology promises for agriculture can be achieved with the science of animals, with the possibility of substantial rewards in animal husbandry, as well as in zoology and preservation of the species, thus ensuring global biodiversity. Balázs et al. give a succinct summary of the benefits of biotechnology to fish farming, a resource exploited almost to exhaustion in many Third World countries, to satisfy markets in developed countries. In their view:

> Biotechnology can play an important role in the elucidation, and subsequent modification, of host genetic factors governing traits of economic interest such as growth rate, stock maturity, efficient food conversion and nutrition, disease resistance, body shape and coloration, temperature tolerance and sex determination.[15]

Some would argue that instead of preserving biodiversity, the genetic manipulation of plant and animal species could in fact result in the extinction of known forms of life and thus deal an irreparable blow to the environment. However, as Richard Oliver states:

> One of the most dramatic insights into molecular biology is the fundamental relationship, the unity of virtually all life at the cellular level. This

> new evidence has sharpened the focus and underscored the need for careful stewardship of biodiversity.[16]

It is precisely because there are varying and sometimes strongly opposed views on such issues, that the time has come for intelligent public debate on the various aspects of the Biotech Age. It is incumbent upon all individuals to inform themselves of the various biotech developments and to ensure that their opinion is heard in influential forums. As Jeremy Rifkin states in his book *The Biotech Century*, "Genetic engineering represents both our fondest hopes and aspirations as well as our darkest fears and misgivings."[17] It is clear that there is an urgent need for discussion and debate on how we can fulfill our aspirations without realizing our darkest fears.

One of the principal features of the new age is that by manipulating the most basic form of life, the cell, and the most basic construct of matter, the atom, new materials can be developed that will change every aspect of the life of people, from their healthcare to the food they eat, and from the materials they use to the environment they live in. It will change the way they earn a living and the way they conduct wars and defense.

In the past, scientists developed modified materials from existing or given materials, but with the new technologies, scientists today can create new materials by modifying their atomic compositions. In other words, scientists customize the atomic architecture of a material to match a predetermined need. John Pierce supports this view in his chapter "Biotechnology and our Material Future." He indicates that biology has become a source of inspiration for materials production. The aim is to find ways to structure existing substances in different ways. Such an approach to materials production will reduce the strain on the world's natural resources, and lower the consumption of energy in manufacture. Pierce estimates that, in another two decades, with cheaper raw material prices, new and cheaper enzymes and production of new materials of higher value at lower costs, biology and biotechnology will have replaced engineering and chemical production. This opens a whole new area of materials research and production, resulting in greater promise of sustainable development, with potential benefits for the environment and for the whole of humanity.

The term "bioterials" was coined by Richard W. Oliver as a way to refer to the commercial applications of biotechnology. It blurs the distinction between organic and inorganic matter, creating a new class of hybrid matter. As Oliver puts it, the Industrial Age harnessed energy, the Information Age harnessed the power of networks and the Bioterials Age is the age of endless replication, a new economic paradigm.[18] The bioterials aspect of the new age frequently fails to receive the prominence it deserves in discussions about biotechnology. Yet, it is an area of great interest to researchers and investors, as the commercial application of new innovations provides the funds for new investment and new research. These commercial applications will not only influence the products available for the consumers on the market, but will also shape the way these products are made. In short, biotechnology is set to influence the way industry approaches manufacture.

A development that is revolutionizing computer technology is the manufacture of a "molecular computer." Rifkin describes it as "a thinking machine of DNA strands rather than silicon."[19] This not only sets the seal on the marriage of biology and computers, but also adds a new dimension to bioinformatics. Rifkin maintains that soon much computing will take place along DNA pathways rather than through the now traditional microchip.[20] He describes DNA as a massive parallel computing machine that can "theoretically compute a hundred million billion things at once."[21] Richard Oliver confirms the importance of the molecular computer in the new Biotech Age and states that although microchip computers work at lightning speed, they "are much too slow for this new world."[22]

One of the more evident manifestations of the impact of biotechnology on industry is in the field of pharmaceuticals. Allan Haberman, in his chapter "Biotechnology and the Future of the Pharmaceutical Industry," states that the global pharmaceutical industry faces a tremendous challenge. Life scientists have provided a host of new technologies and scientific discoveries with high potential for developing breakthrough treatments and cures. There are now big biotech competitors vying with the large multinational pharmaceutical companies, who have until now based their production and profits on the chemical manufacture of drugs. This form of pharmaceutical manufacture has now reached a plateau due to high costs of research and production and the expiry of patents on profitable drugs. In response, large pharmaceutical companies have had to adopt a number of strategies to accommodate the innovations in the field of healthcare, from restructuring their research and development programs to buying out or partnering with biotech research companies. If the pharmaceutical industry is to secure its future, it must have a strategy to utilize new biotechnologies effectively to develop new drugs, realize scientific opportunities and meet social needs.

The changes for the industrial sector inherent in the Biotech Age will impact on the job market, on the use of energy, on the management of industrial waste, on the method and volume of production, and directly on profits that can be made from the application of these technologies. For example, in the area of industrial waste, the potential lies in genetically modified bacterial agents that can render the waste harmless, and may even transform it into commercially useful materials. Smart materials that can modify their behavior over time, learning from previous failures, are also within the realms of feasibility.[23] There is clear evidence of an international competitive environment based on the advances already made in research and in the commercialization of biotechnology.

There is no doubt that biotechnology is going to create global issues of enormous complexity and importance, which will have to receive urgent attention from national governments and international bodies. The ramifications are principally in the areas of trade, public welfare and the environment. In many cases, legislative systems are unable to cope with the implications of the biotechnology revolution. A simple example is the inability of the standard birth certificate and other identification documents to cope with the details of a cloned human being. The legal standing of cloned offspring, should any come into being, is a gray area in law, and illustrates the impact of the advances in biological sciences on various structures of the state.

Another area of both national and international relevance is the issue of patents on the innovations emanating from the biotechnology revolution. Such patents are not only a subject of commercial debate, but the matter of patenting stem cells and genetic sequencing, for example, also has deep ethical significance which requires both an informed and sensitive approach by national governments and international institutions. McGee maintains that the legal response to biotechnological advances is increasingly relevant to define the frameworks for scientific research and determine the permitted scope of biotechnological applications. In this regard, Stock explains that as technology becomes more powerful, in the field of medicine for example, it will face the challenge of how to serve humanity and how to be integrated with existing knowledge and human practices. Currently, stem cell research is operating in a legal twilight zone. Therefore, it receives little government funding, making private companies the main source of funds. This could perhaps lead to a concentration of intellectual property rights in the hands of a small group of companies, raising ethical questions and creating problems of access to biotechnological advances. The issue of bioethics is a contentious subject, generating heated debate on how to determine and ensure ethical conduct on the part of all involved in the unfolding of the new biological era.

Jeremy Rifkin in his chapter on "What Biotechnology Means for the Future of Humanity," also raises the concern that reducing the world's gene pool to patented intellectual property will give patent owners the unprecedented power to dictate the terms by which current and future generations will live. At present, genes are legally considered the intellectual property of those who discover them. This runs counter to the concept of the intrinsic value of life and accords it a utility value. Furthermore, as McGee mentions, other legal issues that emerge from the opportunities and uses of biotechnology are the confidentiality of health information, genetic screening of criminals and civil servants, and genetic discrimination. Such matters have a strong ethical dimension.

The ethics of genetic discrimination is an issue that has a widespread impact. There are concerns that this discrimination will even filter down to education, opening the door for institutions to accept or reject students on the grounds of their genetic construct. The issue of genetic discrimination arises most often in the areas of employment and insurance. As Oliver states, "The concern is that employers and insurers might base decisions on genetic tests that may tell only part of the story."[24] These are issues of public interest that elicit concern. However, Oliver also refers to Craig Venter of Celera Genomics Corporation, who believes that "the amount of genetic information on a single individual is so staggering that it can never be seriously used."[25]

Biotechnology is also certain to have pronounced effects on the relations among nations and, therefore, will influence interstate activity.[26] Some of the more evident areas are technologies applicable to national defense, and technologies of substantial commercial value that may create strong international competition in the world economy, and tension around patent issues. Less evident are the patent issues involving, for example, plant germplasm from Third World countries being used by

First World scientists for biotech innovation. These are matters that have a direct bearing on international law, both commercial and patent laws, and need to be regulated internationally by appropriate institutions, and nationally by responsible state structures. The power of the profit motive must not be underestimated. No technology is of itself good or bad, but the power that it ascribes to the proprietor needs to be circumscribed for the benefit of all.

The impact of genetically altered organisms on the global environment is also a subject of international concern, with legal, social and environmental implications. The study of genomes, which is known as genomics, involves the mapping, sequencing and analysis of genomes in order to determine the structure and function of every gene in an organism. Science has only scratched the surface of this field of endeavor. It is estimated that less than 1 percent of singular-cell organisms have been genetically documented, not to mention multicellular organisms.[27]

At the early stages of the genomic quest, Rifkin makes the statement that it is not yet possible to predict how genetically modified organisms will interact with the ecosystem into which they will be released. International boundaries cannot contain the spread of genetically engineered microorganisms or waste from biotech production processes. In the words of Dale and Irwin as cited in Balázs et al., "The new opportunities to modify plants in novel ways with genetic modification present new responsibilities for safe use to avoid adverse effects on human health and the environment."[28] Balázs et al. add that, "Risk assessment studies are an integral part in producing and placing a transgenic variety on the market."[29] As research and commercialization of biotech products advance at a rapid pace, it becomes imperative for national governments and international institutions to create appropriate regulatory mechanisms to effect biosafety assessments of new biotech products and ensure the optimal utilization of the new technologies within strict parameters of safety, ethics and international responsibility.

Although each new technological age has resulted in a greater dispersion of the benefits of that age to a larger number of people than the preceding age, careful planning needs to be done and effective strategies implemented to ensure a more equitable distribution of the potential benefits. It is not difficult to see that as First World countries forge ahead with the life sciences and concomitant biotechnological innovations, Third World countries will find themselves in ever-greater dependence, exacerbating global development fault lines and leading to further deterioration in international stability. Such dependence may also have a detrimental impact on the internal security of less developed nations, leading to intra-state conflict and further suffering.

In the context of national and international security, the advances in biology and biotechnology, combined with the easy diffusion of such knowledge globally, through computers and the Internet, also imply that biotech innovations could be accessible to individuals or states, which may present a threat to national or international peace and stability. In her chapter "Bio-Terrorism and National Security," Sue Bailey points out that bio-warfare has been around for many decades, even centuries. However, she outlines the increasing risks associated with the proliferation of chemical, biological and nuclear agents in a world destabilized economically, politically and militarily. Such a situation increases the potential for these weapons of mass destruction to be used in asymmetrical warfare or for terrorism. However, biotechnology also holds positive spin-offs for security, for example in criminal identification and security clearance. Among the areas that have been explored are genetic fingerprinting and retina scanning, both aspects of the new technology that promise to simplify individual identification.

A brief overview of the areas of biotech innovations shows that the principal areas that will be impacted upon by advances in the life sciences are pollution, famine, disease, energy and waste management. These are aspects of global concern but they have a particularly keen impact on less developed countries. In an attempt to redress the global developmental imbalance and reduce critical fault lines, a number of international organizations, such as the World Health Organization (WHO), the United Nations Food and Agriculture Organization (FAO) and the United Nations Educational, Scientific and Cultural Organization (UNESCO), perceiving the potential benefits in health and agriculture, have devoted considerable attention to biotechnology on behalf of Third World countries. This has included advising policy makers as well as providing education and training to purchase research equipment. As Michio Kaku mentions, biotechnology differs from nuclear technology in that start-up costs are rather insignificant and substantial knowledge is available on the Internet for free. This facilitates access for poorer countries and provides an opportunity for them to get involved in the field of biotechnology. However, the discrepancy in resources between the economically advanced nations and the less developed countries remains an issue of concern, and much more needs to be done for the benefits of the Biotech Age to become truly global and lead to progress for the whole of humanity.

Bioethics debates have become commonplace in certain government, business and academic circles. It is necessary to ensure that such debates penetrate all levels of society for a broad-based consensus on the scope, depth and direction of the biotechnology revolution. All technologies have a dimension of ethical value: there is no neutral technology, just as there is no inevitable innovation. At all times, scientific innovation should be scrutinized to determine the appropriateness and quality of the contribution they will make. Gregory Stock states that profound changes are going to occur, and we are the architects as well as the objects of this revolutionary transition. The Biotech Age is here and its impact is global. The aspects that remain to be determined are its scope, direction and speed. The bioethics debate must be informed, intelligent, thoughtful and broad-based. The global stage is set for discussion and the need for greater definition and clarity on the broader implications of the Biotech Age is pressing.

Section 2

TRENDS AND PROSPECTS

1

Advances in the Field of Biotechnology

Michio Kaku

Back in the 1980s, when the idea of the Human Genome Project was first proposed by a handful of biologists, the overwhelming reaction was negative, with scientists arguing that it would be prohibitively expensive and would consume too much time and resources. Only a handful of genes had been sequenced, at great expense, and many felt that a crash project to sequence the entire human genome would be impractical and would adversely affect funding for other worthwhile projects.

Today, we realize that many of these pessimistic predictions were incorrect, in part because of Moore's Law. The biology of gene sequencing has now been automated, with the power of computers doubling every 18 months and results being shared instantly on the Internet. This is one of the most important factors driving the ever-accelerating pace of biotechnology. This, in turn, has translated into a new Moore's Law for biotechnology: that the number of genes that are sequenced doubles every year. This means that the cost of sequencing a DNA base pair went down from $5 per base pair to a few cents today. Within twenty years, we may have personalized DNA sequencing and also an "encyclopedia of life" in which all major life forms are decoded.

This new Moore's Law, in turn, allows one to make rough predictions about the progress of biotechnology into the next twenty years. Although the predictions mentioned here are inevitably based on incomplete information, they will hopefully serve as a useful guide to make plausible projections for the future.

The projections in this chapter are based on interviews with one hundred and fifty of the world's top scientists, many of them Nobel Laureates or directors of major scientific laboratories, about their conception of the science for the next twenty to fifty years. Many of these predictions are contained in my book: *Visions: How Science Will Revolutionize the 21st Century.* Of course, some may turn out to be erroneous, but the predictions in this article are not mere idle speculation, reflecting rather a fairly accurate description by biotechnology experts about the evolution of their field.

Advances in the Next Five Years

Given the current pace of discovery, one may expect advances within five years in the following areas:

Stem Cells

Stem cells are the "mother of all cells." Stem cells in the embryo eventually differentiate into the two hundred or so tissues that make up a human body. Thus, isolating and growing embryonic stem cells in the laboratory has been a tremendous breakthrough within the last five years. Although only a handful of stem cells have so far been isolated, for example in blood, cartilage, bone, and so forth, in about a hundred cell lines, eventually most of the stem cells making up the human body will be found. This breakthrough holds out the promise of eventually creating a "human body shop," whereby human organs will be replaced as they wear out, get diseased or become injured.

Within five years, this technology will soon be married to "tissue engineering" technology, which employs special "reactors" to grow simple human tissue in the three-dimensional shape of a human organ. By injecting ordinary cells which have their immune functions suppressed into a sponge-like plastic scaffolding (for example, polyglycolic acid) shaped like a human organ, it is possible to grow simple organs out of bone, cartilage and other tissue. Once the plastic scaffolding dissolves by itself, what is left is a realistic replica of a human organ. In this way, artificial skin, noses, heart valves, and even bladders have already been created. Scientists at the Massachusetts Institute of Technology

(MIT) and the Children's Hospital in Boston have pioneered this technology.

When tissue engineering technology is married to stem cell technology, within five years one should be able to grow more complex organs like the liver and the pancreas (which involve only a handful of different tissues). This may have a dramatic effect on liver transplants and also in treating diabetes. More complex organs, because they involve many different types of tissues with very complex geometries, will remain challenges for the future.

Gene Therapy

Altogether, there are about five thousand known human genetic diseases (for example, cystic fibrosis, which affects mainly northern Europeans, sickle cell anemia, which affects mainly Africans, and hemophilia, which devastated many branches of European royalty). In fact, all of us have approximately half a dozen to a dozen genes that are potentially lethal.

In gene therapy, a "bad" gene can be corrected by inserting the "good" gene into a vector (usually a virus that has been rendered harmless) and then infecting the patient with the virus. The virus then multiplies rapidly, injecting the "good" gene into the cells of the patient. Infected cells can also be grown and cultivated outside the body in order to increase their number and then injected back into the body. The ultimate – and still elusive – goal is to infect virtually all the cells of the body to replace the deficient gene.

In 2002, after a number of mistrials (including a fatality at the University of Pennsylvania in 1999), the first success for gene therapy was recorded in Paris, France. At the Hopital Necker-Enfants Malades, four young boys with Severe Combined Immunodeficiency Syndrome (SCIDS, also known as "bubble boy syndrome"), in which children lack a fully functioning immune system, were the first patients to be treated successfully with this method. They are now relatively free of SCIDS and are able to live outside an artificial, sealed environment.[1]

Within the next five years, many scientists expect that this promising result will be extended to perhaps a handful of other genetic diseases, perhaps five to ten. For example, scientists at the University of Washington have had initial success treating muscular dystrophy in laboratory mice by injecting them with modified genes. This is encouraging news for the twelve thousand patients now suffering from the disease in the United States.

In 2001, scientists at the University of California in San Diego treated the first patient with Alzheimer's disease with gene therapy. A 60-year old woman was injected with cells treated with nerve growth factor in hopes of preventing brain cell death, which typifies Alzheimer's disease.[2]

Gene therapy, however, will still remain experimental and not for the general public for several years, but it may revolutionize the way we view medicine, since a surprising number of common diseases have some genetic basis.

Aging

In 1900, in a Western country such as the United States, the average life expectancy was 49 years. With the introduction of better sanitation and higher living standards, about 15 years were added to this number. As a result of antibiotics, vaccinations and surgery, another 10 years was added. In the future, however, this number will only increase incrementally. To go beyond a life expectancy of 80, new genetic and molecular therapies will have to be introduced.[3]

Scientists do not expect a "fountain of youth" to extend the human lifespan within the next few decades, but the foundations will be laid for unraveling the aging process at the molecular level. Currently, genes that can lengthen the lifespan of simple organisms have been isolated at the University of Colorado and the University of California at San Francisco. These relate to organisms like the nematode worm and fruit flies, with names like Age-1, Age-2 and Daf-2. Through selective breeding, scientists at the University of California at Irvine and McGill University have created new strains of worms and fruit flies with extended lifespans.[4] These age genes, when analyzed carefully, seem to regulate the oxidation within cells, which produces errors in the genetic code. Furthermore, about seventy similar age genes in humans have also been isolated by comparing the genes of younger and older individuals and isolating the ones that are mutated as a person ages.

In addition, scientists now realize that the cells in our body have a biological clock that tells them when to die. Skin cells, for example, have a "Hayflick limit" of fifty reproductions before they die. This is because the ends of the chromosomes are like the fuse of a bomb. After every reproduction, the fuse (the telomere) gets shorter and shorter, until it disappears and the chromosome disintegrates. At the Geron Corporation in Menlo Park, California, scientists have used the enzyme telomerase to prevent shrinkage of the telomeres, thereby prolonging the lifespan of human skin cells to hundreds of reproductions, well past their normal lifespan.[5]

Within the next five years, these encouraging results probably will not translate into any breakthrough that can extend the human lifespan, but the foundation will be laid for understanding aging at the molecular level.

Agriculture

Since plants are easier to manipulate than animals, progress in applying this technology to plants and food crops will progress much faster than for animals. Humans have been cloning plants (e.g. in the form of cuttings) and manipulating their genes (by hybridization and breeding) for millennia. Many of the main crops cultivated on earth (e.g. corn) are the product of thousands of years of gene manipulation by humans. Also, compared with animals, plants have a shorter life cycle, fewer genes in general, are easier to handle, and present fewer ethical issues.

Initially, the modification of plant life was mainly concentrated in the food industry, which wanted to develop genetically modified (GM) foods for the Western market. However, this was a dismal failure, and perhaps the GM market in Europe will be lost for a generation as a result.

Within the next five years, all the main commercially viable food crops will have their complete genome read. This, in turn, will make possible the application of this technology for the Third World. Because of relatively low start-up costs, even countries like Cuba have a vigorous and growing biotech industry for agriculture.

Within five years, there is the possibility of developing new strains of "super rice" and "super corn." These crops will require less fertilizer, will grow in arid and hostile environments, and produce a larger yield than ordinary plants. This new green revolution by itself will not be able to feed the growing populations in the Third World, but it may help to alleviate poverty and hunger to a large degree.

Cloning

Cloning has made remarkable strides ever since the birth of Dolly, the first cloned sheep, in 1997. Since then, a wide variety of farm animals have been cloned, even pets. Now, even second-generation cloned animals exist. However, there are also disturbing problems emerging in the field, which make cloning unsuitable for humans, at least in the short term, and will make cloning difficult for years to come.

First, a large number of fetuses must be sacrificed in the process of producing one healthy clone. This is because the methods of cloning are still quite crude and largely hit-or-miss because of the trauma introduced by reactivating the cell reproduction mechanisms.

Second, a large number of genetic defects are introduced by the process of cloning, injuring the health of the animal. Cloned animals may look normal but actually suffer from obesity, premature aging, arthritis, or any number of other medical problems. In analyzing the genes of the cloned animals, scientists at MIT's Whitehead Institute analyzed ten thousand genes of cloned mice and could see quite clearly the large number of genetic mutations introduced by the cloning process. A cloned mouse, for example, may have several hundred flawed genes.[6]

Consequently, in the next five years, cloning will be restricted to animal husbandry, and then used only for exceptional cases. Mass commercialization of this technology for animal husbandry will be limited. The main focus for the next few years will be to rectify these two problems: to reduce the number of unviable fetuses and the number of mutations introduced by cloning.

For these reasons, it would be immoral as well as impractical to apply this technology to humans within this timeframe. Many nations will pass laws banning this technology. However, within five years, it is conceivable that some unscrupulous scientist will attempt the first human cloning.

Medicine

This period will usher in the era of "molecular medicine." The trial-and-error approach used for the last four thousand years in medicine will gradually be replaced by understanding the molecular and genetic nature of disease and the function of the body.

However, sequencing the genes for animals and diseases alone does not mean that we will understand the function of these genes. For decades to come, our ability to sequence the genes of animals and diseases will far outstrip our ability to understand how these genes work. The genome for an organism is like a gigantic dictionary, containing tens of thousand of words with the correct spelling, but without any definitions. This largely blank dictionary by itself is useless. Within the next five years, scientists will have the complete genome of some of the major diseases plaguing humanity, but will not be able to translate this information readily into cures.

For example, malaria is one of the greatest killers on earth, killing one to two million children each year and infecting several hundred million. In 2002, one hundred and fifty scientists were involved in sequencing the malaria microbe (with five thousand three hundred genes) and the mosquito that carries it (with fourteen thousand genes).[7] However, it may take years fully to digest this wealth of new genetic material to isolate the weak spots in malaria and find new cures.

New breakthroughs will also be made against killers like AIDS and cancer, but progress towards a cure will be slow. The AIDS virus has only nine genes, all of which have been carefully sequenced. In fact, the genetic family tree for AIDS has been deciphered, allowing scientists to reasonably conjecture that it started as a cross-over virus from the monkeys in Africa in the form of the simian virus SIDS, perhaps fifty years ago. However, the virus mutates so rapidly that any short-term cure for the disease seems unlikely. Lacking an effective gene correction mechanism, the AIDS virus is a moving target, mutating so fast it can evade any therapy designed against it, perhaps for years or even decades to come.

Similarly, cancer will remain a stubborn disease for many years to come. A number of cancers involve the successive mutations of four or more genes, and the total number of oncogenes and tumor suppressor genes is enormous. However, new ways are being devised to attack cancer at the molecular level, including starving a tumor of its blood supply, using monoclonal antibodies, and even gene therapy. About 50 percent of all common cancers (e.g. lung, breast, colon cancers) involve mutations in just one gene, p-53. This means that by replacing damaged p-53 genes, we might be able to cure 50 percent of all common cancers. Yet, the reality is not so simple. A large number of the gene therapy trials being conducted today involve trying to fix the mutated p-53 gene, with only mixed results. These show that the good p-53 gene is being absorbed into cancer cells, but that is not enough. It must show that all cancer cells are cured in this way, and this has not been achieved.

So, in the next five years, we can expect to see spectacular breakthroughs in understanding the genetic basis for diseases, but progress towards cures for these diseases will be painfully slow.

Social Debate

The main social debates concerning biotechnology in the next five years will center around privacy issues, food safety and stem cells.

Privacy issues will concern, in the short term, the insurance industry. Since insurance rates can rise if an insurance companies finds out that a potential client suffers from a genetic ailment, an increasing number of governments will pass laws preventing insurance companies and employers from discriminating against prospective customers.

Similarly, the controversy around GM foods will continue to prevent the commercialization of such foods in Europe, but probably not the United States or the rest of the world. Most likely, Europeans will shun GM foods for an entire generation. Until the industry can convince Europeans that GM foods are safe, the GM food industry will languish.

GM foods, in fact are a classic case of how the biotech industry made huge self-inflicted mistakes, which ruined a billion-dollar market. GM foods were hastily introduced, without rigorous controls that might have allayed the fears of the public. The public's first exposure to mass-marketed GM foods was unfortunate, since short-term profits were the main criteria for introducing certain GM foods. Monsanto, for example, offered pesticide-resistant seeds to farmers so that it could sell more pesticides to them. This meant profits for Monsanto, because it made money selling pesticides as well as pesticide-resistant seeds to farmers, but there was no direct benefit to the consumer, who only saw more poisons in their food. (This is like Thomas Edison trying to commercialize electricity at the turn of the century by offering the electric chair, and not the light bulb, as the first commercial product.) Not seeing any benefit deriving from GM foods, disturbed by the increase of pesticides in their foods, and alarmed by the lack of rigorous controls, the public confidence in GM foods plummeted.

This also raises the delicate issue of food labeling. Surveys have shown that consumers, if given the choice, will reject GM foods if they are labeled. However, labeling may be essential to combat the problem of allergies. People allergic to certain nuts, for example, may experience adverse allergic effects if they consume agricultural products where the nut gene has been inserted. Without food labeling, eating may be Russian roulette for the consumer who is allergic to certain foods.

Also, the ethics regarding embryonic stem cells will continue to be controversial because it means that human embryos will have to be discarded. Although thousands of embryos lie frozen in liquid nitrogen tanks at fertility clinics waiting to be thrown in the garbage can, certain religious leaders, especially in the United States and in countries with a strong Catholic or anti-abortion tradition, have argued that embryonic stem cell research should be stopped. For example, in the United States, federal funding is restricted, being available only for 78 existing stem cell lines. These will soon be dwarfed by the rapidly proliferating stem cell lines being found at labs around the world. The most active scientists in this growing field may eventually leave those countries that refuse to fund this technology and may move to countries that view it as a tremendous asset rather than a liability.

Lastly, there are legal problems with regard to genes. Who owns them? Is it the people who donated their bodily fluids or the scientists who analyzed the genes? This is already creating legal problems, especially when certain genes extracted from individuals are found to have great commercial value in the marketplace. Eventually, the courts will have to decide how far a person can argue that they own their own genes.

Also, as large corporations patent the major genes involved in producing key medicines and food products, there is the growing danger of placing too much power in the hands of too few individuals and corporations. The common good may suffer if extremely valuable genes become the property of corporations with only self-interest as a motivating factor.

Advances in the Next Ten Years

Within ten years, scientists will have a vast encyclopedia of the genomes of hundreds or perhaps thousands of diseases, plants and animals. The functions of these genes will be tediously and slowly discovered, most likely by computers. Since it is unethical to experiment genetically on humans, perhaps the fastest way to understand the function of these genes will be to explore bio-informatics; that is, the use of computers to search for homologues of human genes within the plant and animal kingdom. Since humans are about 30 percent equivalent to fungi, 70–90 percent equivalent to many farm animals, and 98.4 percent equivalent to chimpanzees, scientists will be able to determine the function of various genes in the plant and animal kingdom and then find the homologues within the human genome. In this way, the entries in the "encyclopedia of life" will gradually be filled out, gene by gene, in the memory of a computer.

Stem Cells

Within a decade, stem cell research should allow scientists to reliably grow or regenerate a few key organs of the body. Within this timeframe, certain simple organs, like the heart and stomach, or muscles might be grown in the laboratory. (More complex organs, like the kidneys or lungs, may require more time.) This, in turn, will have enormous implications for the millions of people who are waiting for organ transplants. Eventually, an industry may grow up to fill the increasing demands of an aging public. For example, growing cartilage, which has already been demonstrated in the laboratory, may become commonplace within a decade, revolutionizing the way in which arthritis is treated, which could have a profound effect on a rapidly aging Western population. Scientists at MIT have even created a "designer gel," made of cartilage cells and hydrogel scaffolding, which can be grown outside the body and then injected directly into arthritic joints to regenerate new cartilage. The new cartilage has many of the same mechanical properties as natural cartilage.[8]

One organ that will be too difficult to grow is the human brain, because of its complexity and because it is impossible to connect it with the millions of neurons in the spinal cord. However, it should be possible to inject brain stem cells directly into the brain or spinal cord. This should allow for new ways to attack Alzheimer's disease, Parkinson's disease, and spinal cord injuries. The person will not automatically be cured, but instead will have to tediously relearn new activities to integrate the new neural tissue into the brain. Preliminary results on the brains of animals are quite encouraging, with new brain cells growing in the brains of mice treated with stem cells, but it may take years to perfect this technology for the human brain.

Gene Therapy

The number of genetic diseases curable by gene therapy may grow to perhaps fifty to a hundred. Gene therapy, in fact, will become a key arsenal in the war against disease. However, the diseases that can be treated by gene therapy will be restricted to those that are triggered by a single gene rather than those that are multigenic. For many years to come, this technology will be limited to genetic diseases that can be traced to a single mutated gene. Due to the enormous difficulty in isolating diseases, which are caused by a combination of genes and interactions with the environment, multigenic diseases (e.g. mental illness) will take decades to isolate and cure.

Over the years, a number of false claims have been made concerning the isolation of the gene for schizophrenia, mainly because it is caused by a subtle combination of environmental cues and a combination of genes. Painstakingly assembling the family tree of people with a certain genetic disease to isolate a single gene has proven to be a laborious and expensive process. Finding the collection of genes that are involved with schizophrenia may prove to be exponentially more difficult.

On this timescale, a modest but growing list of specific cancers should be curable. Most likely, not one but a variety of methods will be devised to treat a spectrum of cancers at the molecular and genetic level. Injecting the correct version of p-53 into cancer tumors has already proven that gene therapy can temporarily stop the growth of tumors. (Yet, even a single cancer cell that is missed by gene therapy may regenerate the entire tumor.) On this timescale, a few of the experimental approaches (choking off the blood supply, injecting p-53, monoclonal antibodies) should bear fruit.

Agriculture

Contrary to the claims of many in the industry, it is unlikely that the biotech revolution will successfully feed the world's hungry populations. The world's population is growing too rapidly and the successes of biotechnology have been too modest. However, new strains will be at are adapted to harsh environments in the Third World and this will help to alleviate but not eliminate famine. Several thousand animals, plants and diseases will have their genome sequenced, which will alter the way farming, agriculture and disease control is conducted.

Plants and animals may also become commercially valuable in their ability to synthesize rare human proteins. By injecting them with certain genes, they may be used to produce large quantities of life-saving chemicals, thus reducing their cost of manufacture. For example, the US Department of Agriculture in 2002 issued 32 field permits to grow new types of drugs in crops such as barley, rice, tobacco and corn. Scientists hope to grow a large number of drugs, such as plant-based insulin, new vaccines for hepatitis B, cholera and diarrhea, and drugs to treat herpes.

Animals may also be used in this way. Scientists at the University of Milan successfully mixed the sperm of pigs with human DNA to transfer a gene called DAF. Pigs fertilized with this treated sperm created offspring with the human gene. The goal of this technique is to create organs for transplant into humans that do not trigger the same rejection mechanisms as natural animal organs.

Cloning

Many of the current technical problems with respect to the cloning of mammals may be gradually eliminated within this timeframe. Attention will turn not simply to cloning copies of animals but also to creating "designer animals" that have enhanced genetic characteristics. Breeders, instead of relying mainly on studs or particular animals with exceptional characteristics, will be able to isolate the genes for these desirable characteristics and then systematically breed them. Animals that produce more milk and meat, or plants that produce more grain in a harsh environment, will become commonplace. This could improve the productivity of farms without having to use antibiotics, which increase drug-resistant diseases.

Medicine

Although scientists already know the genome of certain common diseases, within this timeframe scientists should be able to use this knowledge to cure many of them. This could have great implications for the Third

World, which is plagued by diseases like malaria, cholera and also certain parasitic diseases. Knowledge of the genome means that we can study these germs from the inside out, learn about their defenses, spot their weaknesses, and find new ways to attack them at the molecular and genetic level.

Similarly, parasites are one of the main factors restricting the growth of people living in certain Third World countries, sapping their strength and vitality. Within this timeframe, the complete genome of the major parasitic diseases will be sequenced, making it possible to find numerous points in its life cycle whereby they may be eradicated at the genetic and molecular level.

One large advance will be the gradual replacement of surgery. Surgery involves major rupture of the skin and organs and exposure of the body to billions of pathogens. It also means many weeks of painful recovery, when the body is susceptible to other diseases like pneumonia. The side effects of surgery are sometimes worse than the original disease.

New advances in microminiaturization may replace surgery with non-invasive technologies. Not only will doctors introduce small fiber-optic wires to guide surgery, but they will also be able to inject entire computers into the body to follow the course of the digestive system and perhaps parts of the circulatory system. Already, MEMs (micro-electrical machines) can make entire engines smaller than the tip of a needle, using the same ultraviolet etching technology used to create microchips for the computer industry. Eventually, these tiny machines and sensors will be placed in the body to carry out diagnostic and surgical functions.

Another advance will be in the area of imaging. Already, PET, MRI and CAT scans have given us three-dimensional representations of the human body, as well as images of the living brain as it thinks. Technical improvements in these imaging techniques should revolutionize the field. For example, one should be able to peer into the arteries of the living heart as it pumps, millisecond by millisecond, and determine precisely how much its arteries have been clogged or how diseased they are. This should vastly improve our ability to monitor heart disease, the number-one killer in the West. (At the present time, our knowledge of the working heart is so primitive that a patient may pass all known cardiac tests yet die of a heart attack on leaving the doctor's office.)

Medicines will also be drastically redesigned by this time. Currently, most drugs have some sort of side effect. The one-size-fits-all drug of today causes side effects because each of us differs at the genetic level. Some of these side effects are potentially lethal, which results in million-dollar lawsuits and also prevents promising drugs from entering the market because they cause adverse effects on a tiny fraction of the population. In the future, when we have personalized DNA sequencing, it may be possible to design drugs that have no side effects at all.

Preventive medicine will also be a dominant theme in this era. For example, our health will be silently monitored by "smart toilets" and "smart clothes" that protect our health, monitoring our heart beat, blood pressure and bodily fluids to give us continuous and accurate readings of our health. These smart toilets may even be able to detect trace amounts of proteins emitted by cancer cells while they form colonies of just a few cells, years before a tumor (consisting of billions of cells) is formed. For example, scientists at Cambridge University announced in 2002 that they have developed techniques to rapidly detect the presence of certain molecules called MCMs that are emitted by cancer cells but not ordinary cells. This simple molecular test for cancer can revolutionize screening for bowel, cervical and other types of common cancer, eventually allowing people to know that cancer colonies are growing in their body even before any tumor can be felt by the fingers. Searching manually for tumors may be a thing of the past.

Social Debate

Within ten years, gene testing (for hundreds of key genes) will become relatively commonplace, not just for the criminal justice system, but even for the general public. The original Human Genome Project took fifteen years and consumed over US$5 billion. Now one of the pioneers of that project, J. Craig Venter, estimates that it would cost US$500,000 to completely sequence a person today. Ultimately, he hopes to drive the cost down to US$1,000 per genome.

Although wealthy individuals will be able to get the entire sequence of their thirty thousand genes placed on a CD, the average person will still be able to afford the sequencing of hundreds of major genes. Already, the introduction of the "silicon DNA chip" made by the Affeymetrix Corporation has made possible rapid diagnoses of certain genetic diseases by analyzing their genome when they are placed on silicon chips. Scientists have even developed a simple breath test where, after a person breathes on a pane of glass, a laboratory can detect mutated versions of p-53 and determine whether the person has lung cancer.[9]

The plummeting cost of gene sequencing could create a major social problem, not with "Big Brother" but with "Little Brother." Tabloid newspapers will clamor to get the genome of movie stars; nosy neighbors will want the genome of their friends and rivals; and parents will want the genome of prospective mates for their children. People will have to scrupulously protect flecks of skin, strands of hair or blood stains from winding up in the hands of people they do not trust. Politics can also be affected, as rival candidates try to get the genome of their opponents to analyze them for any hidden diseases. For example, if the genetic profile of John F. Kennedy had been known ahead of time, this might have disqualified him from running for the President of the United States because he had serious undisclosed kidney problems.

As the vast healing powers of biotech become known to the public, the main social debates will center around the cost of these lifesaving therapies and whether only the wealthy will be able to afford them. Initially, the cost of these lifesaving therapies will be expensive, as will the cost of drugs that take advantage of this new technology. Only the relatively wealthy will be able to have therapies involving the replacement of certain costly organs. In particular, given the enormous costs necessary for the production of entire organs, biotech companies will pass the cost onto the consumer, making certain organs extremely expensive.

Within this timeframe, there may be the first ecological disaster involving genetically modified organisms. For example, given the large number of plants being modified to produce new drugs or new strains, it is inevitable that at some point a few of these genetically modified organisms will accidentally escape into the environment. In the United States there are strict laws mandating that plots of land devoted to GM foods be placed a quarter of a mile from other crops. However, this always leaves open the possibility of pollen blowing from one plot onto another. These organisms, in turn, may act in unexpected ways, displacing native species, or contaminating other food crops, causing an expensive recall of certain foods. For example, the firm Aqua Bounty Farms has created a breed of salmon that grows ten times faster than normal salmon. The company took the genes for growth hormone from one fish and injected them into Atlantic salmon. Although this new breed of salmon may help alleviate food problems in certain parts of the world, it also may have unforeseen consequences if these fish escape into the environment, perhaps displacing normal salmon.[10] Since GM organisms cannot be recalled, like a defective automobile, it could have irreversible effects on the environment. How it handles such a public relations disaster will be a test of the maturity of the biotech industry.

Lastly, advances in unraveling the genetics of behavior could create controversy for the criminal justice system and also for religion. Studies on identical twins separated at birth indicate that human behavior is roughly 50 percent genetically programmed and 50 percent environmentally determined. However, the precise genes involved in human behavior are largely unknown. Within a decade, a few of the combinations of genes that help to govern human behavior may be isolated. People charged with crimes may plead their innocence based on the claim that it was the fault of their genes, which made them prone to violence. In the future, the combination of "gay genes" that predispose (but not determine) a person to be gay may be found. Certain gays may then argue that gayness is partially programmed in their genes and hence it is useless and even injurious to try to re-educate them. However, certain fundamentalist groups might then argue that these genes should be eliminated by gene therapy. As more genes involved with human behavior are isolated, the number of social controversies rapidly escalates.

Advances After Twenty Years

Within twenty years, because of Moore's Law, the price of computer chips will drop to perhaps a penny, or the price of scrap paper. The consumer will have about ten thousand times more computer power than today at the same price. Computation, in fact, will be practically free and universal. This, in turn, could have profound implications on public health, since one of the main driving forces behind biotechnology is the computer.

Stem Cells

Within twenty years, we will perhaps consider the "human body shop" to be commonplace. Nobel Laureate Walter Gilbert predicts that, within twenty years' time, perhaps every organ of the body may be grown in the laboratory.[11] As soon as organs become diseased, old or injured, people will have the option of having stem cells regenerate new organs (in much the way that stem cells in certain reptiles allow them to replace missing limbs and tails). As the Western population ages, this could have a significant impact on those suffering from heart disease, diabetes, Alzheimer's and Parkinson's.

Gene Therapy

Perhaps several hundred genetic diseases (of the five thousand known) should be curable within this timeframe via gene therapy. The next major breakthrough in genetics will be the classification of multigenic diseases, which are the most difficult to identify at the genetic level. These diseases include mental illnesses, which afflict 1 percent of the human race (e.g. schizophrenia, bipolar illness) as well as autoimmune diseases (e.g. arthritis, lupus) and diseases affecting aging (e.g. heart disease). Within this timeframe, many of the combinations of genes contributing to these multigenic diseases may be identified. However, cures for them may prove to be elusive, since scientists will have to figure out how these genes interact and how they take cues from the environment.

Within twenty years the first "designer children" should be possible, so parents might be able to choose a handful of characteristics for their children (including basic body size, shape, and even certain mental characteristics like shyness, thrill-seeking, and so forth) Laws may be passed to restrict this technology, but it will be difficult to stop illegal laboratories from providing this service to anxious parents.

Aging

Within twenty years, with many of the genes responsible for the aging process identified, the first experimental trials will be conducted to extend the human lifespan at the genetic level. Since the genome of millions of people will be known, computers will scan the genes of infants, young people, adults and the elderly, looking for genes that are mutated systematically because of the aging process. These elaborate computer searches should be able to isolate the genes involved in the human aging process.

This could result in new therapies to slow down the aging process, for example using certain drugs to slow down the build-up of metabolic errors in the genome of cells, enhancing the body's ability to repair genetic damage, and using gene therapy to "reset" the biological clock of certain tissues and to slow down the oxidation process within cells.

Cloning

Within this timeframe, cloning for animal husbandry should become relatively commonplace, with many of the technical defects in the process rectified. Some people have speculated that this technology may become so advanced that designer animals may be created that are like the chimeras of Greek legend, that is, half animal and half human. However, this is unlikely, even in this timeframe. The transplanting of certain organs (e.g. wings) probably requires transplanting hundreds or thousands of genes in the precise order, with all the correct connections and engineering. This may be beyond the ability of scientists within the present century.

Attention may turn, instead, to cloning long-extinct animals and bringing them back to life via cloning. Unfortunately, DNA degrades with time. For example, many mammoth carcasses have been discovered in Siberia, frozen in the ice or tundra. One specimen was recently found which is nearly intact and about forty thousand years old. However, DNA specimens taken from such extinct animals show that their DNA has been hopelessly fragmented into thousands of tiny pieces.[12]

After the technical problems of cloning are solved, human cloning may become a distinct possibility. Although it will be banned for the most part, it will be hard to regulate this technology as illegal cloning labs proliferate. However, the market for human cloning will probably be small, and hence relatively unimportant. Only certain individuals may wish for a human clone (e.g. wealthy individuals with no heirs, or parents wishing to replace a deceased child). Like test-tube babies, which caused tremendous ethical debate when they were first introduced but are now accepted without further thought, the public will probably get used to the fact that an insignificant fraction of the human race are clones.

Medicine

At present, a doctor obtains information about one's health principally through primitive tests, such as blood tests, ECG, and so forth. However, within twenty years the price of gene sequencing will drop so astronomically that even the general public will have their own complete personal human genome, which will form the basis of every diagnosis of one's health.

When entering the doctor's office, in fact, the first thing the doctor will do is to take a piece of skin, blood or saliva and perhaps insert it into a small machine (a "genalyzer") which will then produce a CD containing all 30,000 human genes. Then the doctor will insert the CD into a computer to analyze the complete set of probabilities for developing heart disease, cancer and hundreds of other diseases. In this way, a doctor will then be able to provide a map of a person's future health.

With regard to diseases, almost all major life forms will have their complete genome available on the Internet. This will change the way we attack diseases. Currently, finding new antibiotics is usually by trial and error, even though this process has been computerized and roboticized. However, being able to read the genome of entire diseases and isolate their defensive mechanisms at the molecular level will make possible "designer antibiotics" that are created in the memory of computers, which will also reduce their costs.

For example, scientists will be able to identify the molecules necessary to dissolve the cell wall of diseases and hence manufacture new antibiotics that can carry out this task. Then, when the bacteria mutates and develops defenses to the antibiotic, scientists will be able to isolate this mutation and find newer ways to neutralize it without trial and error.

However, even within this timeframe, this does not mean that all diseases will be curable. Diseases, in fact, will probably always exist, especially because viruses and bacteria mutate so rapidly, at times evolving millions of times faster than humans. There will be profound successes in the way we treat diseases of all types, but there will always be some diseases that mutate to evade our techniques.

In addition, "new" diseases like Ebola will constantly catch the world by surprise. These diseases may actually be quite ancient, but their radius of infection was small because the population of infected humans was also small. But with the spread of civilization into newer and more diverse environments, old diseases may suddenly spread around the world with the speed of a jet flight. For example, Legionnaire's disease, Lyme disease, toxic shock syndrome and so forth are probably old diseases that found new avenues to spread with the advance of technology and shifting of human populations into new areas.

Social Debate

Within twenty years, costs should be driven down so that the average person should be able to benefit from much of this technology. The "genetic divide" separating the rich and the poor should be gradually narrowed as the price of this technology continues to plunge. This may also follow the way in which the "electric divide" and the "digital divide" evolved. Although electricity (at the turn of the nineteenth century) and PCs (in the 1990s) were initially expensive and created concerns about the division between rich and poor, the price of electricity plunged in the twentieth century, and the price of computer chips will drop within the next twenty years to a penny, making electricity and computing power essentially free. Genetics may also gradually follow this same path.

The main social debate will revolve around how far to take the genetic engineering of the human race, and which characteristics should be considered desirable or undesirable. In the Western world, as its population continues to age, there will be a debate as to how much to extend the human lifespan. If medicine lengthens the human lifespan, but not the quality of life, then the healthcare system of these nations may be overwhelmed as the population ages. The key will be to lengthen the lifespan while also keeping people vigorous and healthy. As the birth rate and population continues to plummet, there could be serious social divisions within these countries, between an aging population and newer immigrants.

In the Third World, nations may suffer the opposite problem. Advances in medicine will continue to lengthen life expectancy in these countries, making it difficult to feed their expanding populations or to give them adequate jobs. This could cause social unrest, as it already has in many areas of the world since the introduction of better sanitation.

Also, in twenty years we will be able to control human evolution to a degree and to decide the genetic heritage of the human species. This will produce perhaps the most profound debate in human history, namely determining how far we should redesign ourselves. For example, if designer children are possible, then some scientists have argued that strict laws should be passed limiting the number and types of genes that can be altered. Some scientists have argued that somatic gene therapy should be allowed for a variety of genetic defects, but germline gene therapy, which permanently alters the genes of our offspring, should be banned. Others have argued that germline gene therapy should be allowed as long as it is used to eliminate disease but not to enhance a person cosmetically, making them more handsome or beautiful. (However, since parents are genetically hardwired to give every evolutionary advantage to their offspring, there will always be a black market of laboratories offering illegal enhancements for people's offspring. Given the importance of being attractive and getting good grades in our society, there will be enormous pressure from parents to obtain this technology for their children.)

Likewise, dictatorships may try to take advantage of this technology. They may try to manufacture truly hideous designer germs, such as airborne AIDS or airborne Ebola that could in principle wipe out hundreds of millions of people. Or they may try to create a genetically programmed army of soldiers with exceptional strength and loyalty but limited intelligence. Even a small dictator may have tools that Hitler only dreamed of.

This technology, in principle, is more powerful than the atomic bomb. However, the nuclear age was ushered in during the Cold War, and hence government secrecy covered up huge disasters and unethical behavior. Today, we see numerous contaminated nuclear waste sites in both the United States and Russia that are like sores on the surface of the earth. The total cost of the clean-up could easily reach US$500 billion.

Today, biotechnology is growing like the nuclear industry in the last century. One difference from the nuclear age, however, is that biotechnology is being developed under the full glare of the media and the scrutiny of critics. In this sense, its ultimate fate will be determined by democratic debate. The more people become familiar with the power and promise of this technology, the more they can engage in mature and sophisticated debates about its potential. The key to the future of biotechnology is to educate people so that they can democratically and wisely determine how far this powerful technology should be developed and applied.

2

What Biotechnology Means for the Future of Humanity

Jeremy Rifkin

Our way of life is likely to be transformed more fundamentally in the next several decades than in the previous one thousand years. By the year 2025, our own generation and the next may be living in a world utterly different from anything that human beings have ever experienced in the past. Long-held assumptions about nature, including our own human nature, are likely to be reconsidered. Many age-old practices regarding sexuality, reproduction, birth and parenthood could be partially abandoned. Ideas about equality and democracy are also likely to be redefined, as well as our vision of what is meant by terms such as "free will" and "progress."

There are many forces converging to create this powerful new social current. At the epicenter is a technology revolution historically unparalleled in its power to remake ourselves, our institutions and our world. Scientists are beginning to reorganize life at the genetic level. The new tools of biology are opening up opportunities for refashioning life on Earth. Before us lies an uncharted new landscape, with its contours being shaped in thousands of biotechnology laboratories in universities, government agencies and corporations around the world. Even if the claims already being made for the new science are only realized partially, the consequences for society and future generations are likely to be enormous. In this chapter, we discuss some examples of the developments that could occur within the next twenty-five years.

A handful of global corporations, research institutions and governments could hold patents on virtually all the one hundred thousand or more genes that make up the blueprint of the human race, as well as the cells, organs and tissues that comprise the human body. They may also own similar patents on tens of thousands of microorganisms, plants and animals, allowing them unprecedented power to dictate the terms by which current and future generations will live their lives.

Global agriculture could find itself in the midst of a great transition in world history, with an increasing volume of food and fiber being grown indoors in tissue culture in giant bacteria baths at a fraction of the price of growing staples on the land. The shift to indoor agriculture could presage the eventual elimination of the agricultural era that stretched from the Neolithic revolution some ten thousand years ago to the green revolution of the latter half of the twentieth century. While indoor agriculture could mean cheaper prices and a more abundant supply of food, millions of farmers in both the developing and developed world could be uprooted from the land, sparking one of the great social upheavals in world history.

Tens of thousands of novel transgenic bacteria, viruses, plants and animals could be released into the Earth's ecosystems for commercial tasks ranging from "bioremediation" to the production of alternative fuels. Some of those releases, however, could wreak havoc with the planet's biosphere, spreading destabilizing and even deadly "genetic pollution" across the world. Military uses of the new technology might have equally devastating effects on the Earth and its inhabitants. Genetically engineered biological warfare agents could pose as serious a threat to global security in the coming century as nuclear weapons do now.

Animal and human cloning could become commonplace, with "replication" partially replacing "reproduction" for the first time in history. Genetically customized and mass-produced animal clones could be used as chemical factories to secrete – in their blood and milk – large volumes of inexpensive chemicals and drugs for human use. We could also see the creation of a range of new chimeric animals on Earth – including human/animal hybrids – to be used as experimental subjects in medical research and as "organ donors" for xenotransplantation. The artificial creation and propagation of cloned, chimeric and transgenic animals could mean the end of the "wild" and the substitution of a "bio-industrial" world.

Some parents might choose to have their children
test tubes and gestated in artificial wombs outside the hu
avoid the unpleasantness of pregnancy and to ensure a safe, ...nsparent environment through which to monitor their unborn child's development. Genetic changes could be made in human fetuses in the womb to correct deadly diseases and disorders and to enhance mood, behavior, intelligence and physical traits. "Customized" babies could pave the way for the rise of a eugenic civilization in the twenty-first century.

Millions of people could obtain a detailed genetic read-out of themselves, allowing them to gaze into their own biological futures. This genetic information would give people the power to predict and plan their lives in ways never before possible. However, the same "genetic information" could be used by schools, employers, insurance companies and governments to determine educational tracks, employment prospects, insurance premiums and security clearances, giving rise to a new and virulent form of discrimination based on one's "genetic profile." Our notions of sociality and equity could be transformed. Meritocracy could give way to "genetocracy," with individuals, ethnic groups and races increasingly categorized and stereotyped by genotype, making way for the emergence of an "informal" biological caste system in countries around the world.

The biotech century could bring some or even most of these changes and many more into our daily lives, deeply affecting our individual and collective consciousness, the future of our civilization, and the biosphere itself. We are in the throes of one of the great transformations in world history. We are moving from the Age of Fossil Fuels and Metals to the Age of Genes.

The Marriage of Computers and Genes

For more than a decade, futurists, economists and policy makers have heralded the emergence of the Information Age while giving short shift to the revolutionary developments occurring in the life sciences. Now that is beginning to change. The two great technological revolutions of the last quarter-century are finally coming together to create a single new economic epoch in world history. All over the world, researchers are using computers to download, catalogue and organize the vast genetic information that is the raw resource of the emerging biotech economy. In government, university and corporate laboratories, researchers are using computers to map and sequence the entire genomes of creatures from the lowliest bacteria to human beings, with the goal of finding new ways of harnessing and exploiting genetic information for economic purposes. By the end of the twenty-first century, molecular biologists hope to have downloaded and catalogued the genomes of thousands of organisms – a vast library containing the evolutionary "blueprints" of many of the microorganisms, plants and animals that populate the Earth. Mapping the genomes of so many species "will yield quantities of information that will dwarf by orders of magnitude anything encountered before," says biochemist Charles Cantor, the chief scientist at the Department of Energy's human genome project. The biological information being generated is so great that it can only be managed by computers and stored electronically in thousands of databases around the world.

Mapping and sequencing the genomes is just the beginning. Understanding and chronicling all of the webs of relationships between genes, tissues, organs, organisms and external environments, and the perturbations that trigger genetic mutations and phenotypical responses, is so far beyond any kind of complex system ever modeled or deciphered that only an interdisciplinary approach, leaning heavily on the computational skills of the information scientists, can hope to accomplish the task.

Not surprisingly, bioinformatics, once a backwater of molecular biology, has suddenly come of age. Titans in the computer field like Bill Gates, and Wall Street insiders like Michael Milken, are pouring funds into the new field of bioinformatics, in hopes of advancing the marital partnership of the information and life sciences.

Patenting Life

The mapping of the human genome has focused public attention on a revolutionary change that is taking place in the global economy. We are in the midst of a historic transition from the Industrial Age to the Biotech Age. While the twentieth century was shaped largely by spectacular breakthroughs in physics and chemistry, the twenty-first century will belong to the biological sciences.

Genes are the raw resource of the new economic epoch. Molecular biologists around the world are mapping the genomes of many of the

Earth's creatures, from the lowliest bacteria to human beings, creating a vast genetic library for commercial exploitation. Gene technology is already being employed in a variety of business fields – including agriculture, animal husbandry, energy, construction materials, pharmaceuticals, medicine and food and drink – to fashion a bioindustrial world.

At the heart of any public discussion of the new genetic commerce is the issue of patenting the genetic blueprints of millions of years of evolution. The economic and political forces that control the genetic resources of the planet will exercise tremendous power over the future world economy, just as in the Industrial Age when access to and control over fossil fuels and valuable metals and minerals helped to determine control over world markets.

In the years ahead, the planet's shrinking gene pool is going to become a source of increasing monetary value. Multinational corporations are already scouting the continents to locate microbes, plants, animals and humans with rare genetic traits that might have market potential. After locating the desired traits, biotech companies are modifying them and seeking patent protection for their new "inventions."

Corporate efforts to "commodify" the gene pool are meeting with strong resistance from a growing number of nongovernmental organizations and countries in the Southern Hemisphere, which are beginning to demand an equitable sharing of the fruits of the biotech revolution. While the technological expertise needed to manipulate the new "green gold" resides in scientific laboratories and corporate boardrooms in the North, most of the genetic resources needed to fuel the new revolution lie in the ecosystems of the South.

Extending patents to life raises the important legal question of whether engineered genes, cells, tissues, organs and whole organisms are truly human inventions or merely discoveries of nature that have been skillfully modified by human beings. In order to qualify as a patented invention in most countries, the inventor must prove that the object is novel, non-obvious and useful – that is, that no one has ever made the object before, that the object is not something that is so obvious that someone might have thought of it using prior art, and that the object serves some useful purpose. Yet, even if something is novel, non-obvious and useful, if it is a discovery of nature it is not an invention and, therefore, not patentable. For this reason, the discovery of chemical elements in the periodic table, while unique, non-obvious when first isolated and purified, and very useful, were none the less *not* considered patentable as they were discoveries of nature, despite the fact that some degree of human ingenuity went into isolating and classifying them. The United States Patent Office (PTO) has said, however, that the isolation and classification of a gene's properties and purposes is sufficient to claim it as an invention.

The prevailing logic becomes even more strained when consideration turns to patenting a cell, or genetically modified organ, or whole animal. Will a kidney or pancreas become patentable simply because it has been subjected to a slight genetic modification? What about a chimpanzee? This is an animal that shares 99 percent of the genetic make-up of a human being. Should it qualify as a human invention if researchers insert a single gene into its biological make-up? The answer, according to the patent office, is affirmative.

The patent issue is likely to become a question of increasing public concern as a result of the stunning breakthroughs in the government-funded Human Genome Project. It is expected that in less than ten years, nearly all of the genes that make up the genetic blueprints of the human race will have been identified and become the intellectual property of transnational life science companies. Such firms are also patenting human chromosomes, cell lines, tissues and organs. PPL Therapeutics, the life science company that cloned the sheep named Dolly, has received a patent from the British patent office that includes cloned human embryos as intellectual property.

The increasing consolidation of corporate control over the genetic blueprints of line, as well as the technologies to exploit them, is alarming because the biotech revolution will affect every aspect of our lives. The way we eat, the way we date, the way we have babies, the way we raise and educate children, the way we work, even the way we perceive the world around us and our place in it; all of our individual and shared realities will be deeply touched by the biotech revolution.

Life patents strike at our core beliefs about the very nature of life and whether it is to be conceived of as having intrinsic or mere utility value. The last great debate of this kind occurred in the nineteenth century over the issue of human slavery, with abolitionists arguing that every human being has "God-given rights" and cannot be made the commercial property of another. Like anti-slavery abolitionists, a new generation of genetic activists is beginning to challenge the concept of patenting human life, arguing that human genes, chromosomes, cell lines, tissues, organs and embryos should not be reduced to commercial intellectual property controlled by global conglomerates and traded as mere utilities.

The mapping of the Human Genome ought to be regarded as a triumph for the whole of the human race. Similarly, the knowledge that will come from locating all the genes that make up our common biological destiny should be a shared human responsibility.

The battle to keep the Earth's gene pool an open commons, free of commercial exploitation, will be one of the critical struggles of the biotech century. "Genetic rights," in turn, is likely to emerge as the seminal issue of the coming era, defining much of the political agenda in the years ahead.

A Second Genesis and the Spread of Gene Pollution

The globalization of commerce and trade makes possible the wholesale reseeding of the Earth's biosphere with a laboratory-conceived second Genesis, an artificially produced bio-industrial nature designed to replace nature's own evolutionary scheme. The growing arsenal of biotechnologies is providing us with powerful new tools to engage in what will likely be the most radical human experiment on the Earth's life forms and ecosystems in history. Imagine the wholesale transfer of genes between totally unrelated species and across all biological boundaries – plant, animal and human – creating thousands of novel life forms in a brief moment of evolutionary time. Then, with clonal propagation, mass-producing countless replicas of these new creations, releasing them into the biosphere to propagate, mutate, proliferate and migrate, colonizing the land, water and air. This great scientific and commercial experiment is, in fact, underway as we turn the corner into the Biotech Century.

A global life science industry is already beginning to wield unprecedented power over the vast biological resources of the planet. Life science fields ranging from agriculture to human medicine are being brought under the umbrella of giant "life" companies in the emerging biotech marketplace. The consolidation of the life sciences industry by global commercial enterprises rivals the consolidations, mergers and acquisitions going on in the other great technology arena of the twenty-first century, computer telecommunications, entertainment and information services, although much less attention has been focused on the life companies in the media and public policy.

The concentration of power is impressive. The top ten agrochemical companies control 81 percent of the US$29 billion global agrochemical market. Ten life science companies control 37 percent of the US$15 billion per year global seed market. The world's ten major pharmaceutical companies control 47 percent of the US$197 billion pharmaceutical market. Ten global firms now control 43 percent of the US$15 billion veterinary pharmaceutical trade. Topping the life science list are ten transnational food and drink companies whose combined sales exceeded US$211 billion in 1995.

Corporate leaders in the new life sciences industry promise not only an embarrassing level of riches but also open the door to a new era of history where evolution itself becomes subject to human authorship. Critics worry that the reseeding of the Earth with a second Genesis could lead to a far different future involving the spread of "genetic pollution" throughout the biological world in the coming century, destroying habitats, destabilizing ecosystems, and diminishing the remaining reservoirs of biological diversity on the planet.

Global life-science companies are expected to introduce thousands of new genetically engineered organisms into the environment in the coming century. In just the past three years, genetically engineered corn, soy and cotton have been planted over millions of acres of US farmland. Genetically engineered insects, fish and domesticated animals have also been introduced.

Virtually every genetically engineered organism released into the environment poses a potential threat to the ecosystem. To appreciate why this is so, we need to understand why the pollution generated by genetically modified organisms is so different from the pollution resulting from the release of petrochemical products into the environment.

Genetically engineered organisms differ from petrochemical products in several important ways. Because they are alive, genetically engineered organisms are inherently more unpredictable than petrochemicals in the way they interact with other living things in the environment. Consequently, it is much more difficult to assess all of the potential effects that a genetically engineered organism might have on the Earth's ecosystems.

Genetically engineered products also reproduce. They grow and they migrate. Unlike petrochemical products, it is difficult to constrain them within a given geographical locale. Finally, once released, it is virtually impossible to recall genetically engineered organisms back to the laboratory, especially those organisms that are microscopic in nature. For all these reasons, genetically engineered organisms may pose far greater long-term potential risks to the environment than petrochemical substances.

The risks in releasing novel genetically engineered organisms into the biosphere are similar to those encountered in introducing exotic organisms into the North American habitat. Over the past several hundred years, thousands of non-native organisms have been brought to America from other regions of the world. While many of these creatures have adapted to the North American ecosystems without severe dislocations, a small percentage of them have run wild, wreaking havoc on the flora and fauna of the continent. The Gypsy moth, Kudzu vine, Dutch elm disease, chestnut blight, starlings, and Mediterranean fruit flies are examples that come easily to mind.

Whenever a genetically engineered organism is released, there is always a small chance that it too will run amok because, like non-indigenous species, it has been artificially introduced into a complex environment that has developed a web of highly integrated relationships over long periods of evolutionary history. Each new synthetic introduction is tantamount to playing ecological roulette. That is, while there is only a small chance of it triggering an environmental explosion, if it does, the consequences could be significant and irreversible.

The insurance industry revealed several years ago that it would not insure the release of genetically engineered organisms into the environment against the possibility of catastrophic environmental damage because the industry lacks a risk-assessment science – a predictive ecology – with which to judge the risk of any given introduction. In short, the insurance industry clearly understands the Kafkaesque implications of a government regime claiming to regulate a technology in the absence of clear scientific knowledge of how genetically modified organisms interact, once introduced into the environment. Increasingly nervous over the insurance question, one of the biotech trade associations attempted to raise an insurance pool among its member organizations, but gave up when it failed to raise sufficient funds to make the pool operable. Some observers worried then, and continue to worry – albeit privately – over what might happen to the biotech industry if a large-scale commercial release of a genetically altered organism resulted in a catastrophic environmental event. For example, the introduction and spread of a new weed or pest comparable to the Kudzu vine, Dutch elm disease or gypsy moth, which might inflict costly damage to flora and fauna over extended eco-ranges. Corporate assurances aside, one or more significant environmental mishaps are an inevitability in the years ahead. When that happens, every nation is going to be forced to address the issue of liability. Farmers, landowners, consumers and the public at large are going to demand to know how it could have happened and who is liable for the damages inflicted. When the day arrives – probably sooner rather than later – "genetic pollution" will take its place alongside petrochemical and nuclear pollution as another grave threat to the Earth's already beleaguered environment.

Nowhere are the alarm bells going off faster than in agricultural biotechnology. The life science companies are introducing biotech crops into the field containing novel genetic traits from other plants, viruses, bacteria and animals. The new genetically engineered crops are designed to perform in ways that have eluded scientists working with classical breeding techniques. Scientists have inserted "antifreeze" protein genes from flounder into the genetic code of tomatoes to protect the fruit from frost damage. Chicken genes have been inserted into potatoes to increase disease resistance. Firefly genes have been injected into the biological code of corn plants to serve as genetic markers. Chinese hamster genes have been inserted into the genome of tobacco plants to increase sterol production.

Ecologists are unsure of the consequences of bypassing natural species boundaries by introducing genes into crops from wholly unrelated plant and animal species. The fact is, there is no precedent in history for this kind of "shotgun" experimentation. For more than ten thousand years, classical breeding techniques have been limited to the transference of genes between closely related plants or animals that can sexually interbreed, limiting the number of possible genetic combinations. Natural evolution appears to be similarly circumscribed. By contrast, the new gene-splicing technologies allow us to bypass all previous biological boundaries in nature, creating life forms that have never before existed. For example, consider the ambitious plans to engineer transgenic plants to serve as pharmaceutical factories for the production of chemicals and drugs. Foraging animals, seed-eating birds, and soil insects will be exposed to a range of genetically engineered drugs, vaccines, vitamins, industrial enzymes, plastics and hundreds of other foreign substances, for the first time, with untold consequences. The notion of large numbers of species consuming plants and plant debris containing a wide assortment of chemicals that they would normally never be exposed to is an unsettling prospect.

Much of the current effort in agricultural biotechnology is centered on the creation of herbicide-tolerant, pest-resistant and virus-resistant plants. Herbicide-tolerant crops are a favorite of companies like Monsanto and Novartis who are anxious to corner the lucrative worldwide market for their herbicide products. More than 600 million pounds of poisonous herbicides are dumped on US farmland each year, most sprayed on corn, cotton and soybean crops. Chemical companies gross more than US$4 billion dollars per year in US herbicide sales alone.

To increase their share of the growing global market for herbicides, life science companies have created transgenic crops that tolerate their own herbicides. The idea is to sell farmers patented seeds that are resistant to a particular brand of herbicide in the hope of increasing their share of both the seed and herbicide markets. Monsanto's new herbicide-resistant patented seeds, for example, are resistant to its best-selling chemical herbicide, Roundup.

The chemical companies hope to convince farmers that the new herbicide-tolerant crops will allow for a more efficient eradication of weeds. Farmers will be able to spray at any time during the growing season, killing weeds without killing their crops. Critics warn that with new herbicide-tolerant crops planted in the fields, farmers are likely to use even greater quantities of herbicides to control weeds, as there will be less fear of damaging their crops in the process of spraying. The increased use of herbicides, in turn, raises the possibility of weeds developing resistance, forcing an even greater use of herbicides to control the more resistant strains.

The new generation of virus-resistant transgenic crops pose the equally dangerous possibility of creating new viruses that have never before existed in nature. Concerns are surfacing among scientists and in the scientific literature over the possibility that the coat-protein genes could recombine with genes in related viruses that find their way naturally into the transgenic plant, creating a recombinant virus with novel features. The prospect of creating new viruses is troubling and raises serious doubts as to the safety and efficacy of releasing virus-resistant transgenic crops into the open environment.

A growing number of ecologists warn that the biggest danger might lie in what is called "gene flow" – the transfer of transgenic genes from crops to weedy relatives by way of cross-pollination. Researchers are concerned that transgenic genes for herbicide tolerance, and pest and viral resistance, might escape and, through cross-pollination, insert themselves into the genomes of weedy relatives, creating weeds that are resistance to herbicides, pests and viruses. Fears over the possibility of transgenic genes jumping to wild weedy relatives heightened in 1996 when a Danish research team, working under the auspices of Denmark's Environmental Science and Technology Department, observed the transfer of a transgene from a transgenic crop to the genome of a wild weedy relative. Critics of deliberate release experiments have warned of such a gene flow for years but biotech companies have dismissed it as a remote or nonexistent possibility.

Transnational life science companies project that within less than ten to fifteen years, all of the major crops grown in the world will be genetically engineered to include herbicide-, pest-, virus-, bacterial-, fungus- and stress-resistant genes. Millions of acres of agricultural land and commercial forest will be transformed in the most daring experiment ever undertaken to remake the biological world. Proponents of the new science, armed with powerful gene-splicing tools and precious little data on potential impacts, are charging into this new world of agricultural biotechnology, giddy over the potential benefits and confident that the risks are minimum or nonexistent. They may be right. However, what if they are wrong? What might the consequences be of unleashing genes into the biosphere that are resistant to herbicides, pests, viruses, bacteria, fungus and stress?

The introduction of novel genetically engineered organisms also raises a number of serious human health issues as yet unresolved. Health professionals and consumer organizations are most concerned about the potential allergenic effects of genetically engineered foods. The FDA announced in 1992 that special labeling for genetically engineered foods would not be required, touching off protest among food professionals, including the nation's leading chefs and many wholesalers and retailers.[1]

With 2 percent of adults and 8 percent of children ha
responses to commonly eaten foods, consumer advocates
gene-spliced foods need to be properly labeled so that consumers can avoid health risks. Their concerns were heightened in 1996 when the *New England Journal of Medicine* published a study showing genetically engineered soybeans containing a gene from a Brazil nut could create an allergic reaction in people who were allergic to the nuts.[2] The test result was unwelcome news for Pioneer Hi-Bred International, the Iowa-based seed company that hoped to market the new genetically engineered soy. Though the FDA said it would label any genetically engineered foods containing genes from common allergenic organisms, the agency fell well short of requiring across-the-board labeling. The *New England Journal of Medicine* editors questioned what kind of protection consumers would have against genes from organisms that have never before been part of the human diet and that might be potential allergens. Concerned over the agency's seeming disregard for human health, the *Journal* editors concluded that FDA policy "would appear to favor industry over consumer protection."

Most molecular biologists and the biotechnology industry at large have all but dismissed the growing criticism of ecologists, whose recent studies suggest that the biotech revolution will likely be accompanied by the proliferation and spread of genetic pollution and the wholesale loss of genetic diversity. Nonetheless, the uncontrollable spread of "super weeds," the build-up of resistant strains of bacteria and new "super insects," the creation of novel viruses, the destabilization of whole ecosystems, the genetic contamination of food, and the steady depletion of the gene pool are no longer minor considerations or the mere grumbling of a few disgruntled critics. To ignore such warnings is to place the biosphere and civilization in harm's way in the coming years. Pestilence, famine and the spread of new kinds of diseases throughout the world might yet turn out to be the final act in the script being prepared for the Biotech Century.

A Eugenic Civilization

When Aldous Huxley wrote his dystopian novel, *Brave New World*, in 1932, neither he nor his contemporaries could have imagined that by the end of the twentieth century, the scientific insights and technological know-how would be available to make his vision of a eugenic civilization a reality.[3] The prospect of creating a new eugenic man and woman is becoming ever more likely as a result of the extraordinary advances occurring in the field of genetic technology.

Over the next ten years, molecular biologists say they will locate specific genes associated with several thousand genetic diseases. In the past, a parent's genetic history provided some clues to genetic inheritance, but there was still no way to know for sure whether specific genetic traits would be passed on. In the future, the guesswork will be increasingly eliminated, posing a moral dilemma for prospective parents. Parents will have at their disposal an increasingly accurate read-out of their individual genetic make-ups, and will be able to predict the statistical probability of a specific genetic disorder being passed on to their children as a result of their biological union.

To avoid the emotional anguish of such decisions, some young people are likely to opt for prevention and avoid marrying someone of the wrong "genotype" for fear of passing on serious genetic diseases to their offspring. Already, part of the orthodox Jewish community in the United States has established a nationwide program to screen all young Jewish men and women for Tay-Sachs disease. Every young Jew is encouraged to take the test. The results are made available in an easily accessible database to allow young eligible men and women to choose their dating partners with genotype in mind.

While genetic screening is already here, human genetic engineering – gene therapy – is just around the corner. Genetic manipulation is of two kinds. In somatic therapy, intervention takes place only within somatic cells and the genetic changes do not transfer into the offspring. In germline therapy, genetic changes are made in the sperm, egg or embryonic cells, and are passed along to future generations. Somatic gene surgery has been carried out in limited human clinical trials for more than seven years. Germline experiments have been successfully carried out on mammals for more than a decade and researchers expect the first human trials to be conducted within the next several years.

Programming genetic changes into the human germ line to direct the evolutionary development of future generations is the most radical human experiment ever contemplated and raises unprecedented moral, social and environmental risks for the whole of humanity. Even so, a growing number of molecular biologists, medical practitioners and pharmaceutical companies are anxious to take the gamble, convinced that controlling our evolutionary destiny is humankind's next great social frontier. Their arguments are couched in terms of personal health, individual choice and collective responsibility for future generations.

Writing in the *Journal of Medicine and Philosophy*, Dr Burke Zimmerman makes several points in defense of germline cell therapy over somatic cell therapy. To begin with, he argues that the increasing use of somatic therapy is only likely to increase the number of survivors with defective genes in their germ lines – genes that will continue to accumulate and further "pollute" the genetic pool of the species, passing an increasing number of genetic problems onto succeeding generations. Secondly, although somatic therapy may be able to treat many disorders in which treatment lies in replacing populations of cells, it might never prove effective in addressing diseases involving solid tissues, organs, and functions dependent on structure – for example the brain – and therefore, germline therapy is likely to be the only remedy, short of abortion, against such disorders.[4]

Zimmerman and other proponents of germline therapy argue for a broadening of the ethical mandate of the healing professions to include responsibility for the health of those not yet conceived. The interests of the patient, they say, should be extended to include the interests of "the entire genetic legacy that may result from intervention in the germ line." Moreover, parents ought not to be denied their right as parents to make choices on how best to protect the health of their unborn children during pregnancy. To deny them the opportunity to take corrective action in the sex cells or at the early embryonic stage would be a serious breach of medical responsibility. Proponents of germline therapy ask why millions of individuals need to be subjected to painful, intrusive and potentially risky somatic therapy when the gene or genes responsible for their diseases could be more easily eliminated from the germ line, at less expense and with less discomfort.

Finally, the health costs to society need to be factored into the equation, say the advocates of germline therapy. Although the costs of genetic intervention into the germ line to cure diseases are likely to remain high in the early years, the cost is likely to drop dramatically in the future as the methods and techniques become more refined. The lifetime cost of caring for generations of patients suffering from Parkinson's disease or severe Down's syndrome is likely to be far greater than simple prevention in the form of genetic intervention at the germline level.

In the coming decades, scientists will learn more about how genes function. They will become increasingly adept at turning genes "on" and "off." They will become more sophisticated in the techniques of recombining genes and altering genetic codes. At every step, conscious decisions will have to be made as to which kinds of permanent changes in the biological codes of life are worth pursuing and which are not. A society and civilization steeped in "engineering" the gene pool of the planet cannot possibly hope to escape the kind of ongoing eugenics decisions that go hand and hand with each new advance in biotechnology. There will be enormous social pressure to conform with the underlying logic of genetic engineering, especially when it comes to its human applications.

Parents in the biotech century will be increasingly forced to decide whether to take their chances with the traditional genetic lottery and use their own unaltered egg and sperm, knowing their children may inherit some "undesirable" traits; or undergo corrective gene changes on their sperm, egg, embryo or fetus; or substitute egg or sperm from a donor through in vitro fertilization and surrogacy arrangements. If they choose to go with the traditional approach and let genetic fate determine their child's biological destiny, they could find themselves culpable if something goes dreadfully wrong in the developing fetus, something they could have avoided had they availed themselves of corrective genetic intervention at the sex cell or embryo stage.

Proponents of human genetic engineering argue that it would be cruel and irresponsible not to use this powerful new technology to eliminate serious "genetic disorders." The problem with this argument, says the *New York Times* in an editorial entitled, "Whether to Make Perfect Humans," is that "there is no discernible line to be drawn between making inheritable repair of genetic defects and improving the species." The *Times* rightly points out that once scientists are able to repair genetic defects, "it will become much harder to argue against additional genes that confer desired qualities, like better health, looks or brains."[5]

If diabetes, sickle cell anemia and cancer are to be prevented by altering the genetic make-up of individuals, why not proceed to other less serious "defects": myopia, color blindness, dyslexia, obesity, left-handedness? Indeed, what is to preclude a society from deciding that a certain skin color is a disorder? In the end, why would we ever say no to any alteration of the genetic code that might enhance the well-being of our offspring? It would be difficult to imagine parents rejecting any genetic modifications that promised to improve the opportunities for their progeny.

With Americans already spending billions of dollars on cosmetic surgery to improve their looks and on psychotropic drugs to alter their mood and behavior, the use of genetic therapies to enhance their unborn children also seems a likely prospect. According to a Harris poll, 43 percent of Americans "would approve using gene therapy to improve babies' physical characteristics." Many advocates of germline intervention are already arguing for enhancement therapy. They contend that the current debate over corrective measures to address serious illnesses is too limited and urge a more expansive discussion to include the advantage of enhancement therapy as well. The oft-heard criticism is that genetic enhancement will favor children of the rich at the expense of children of the poor – as the rich will be the only ones capable of paying for genetic enhancement of their offspring. Proponents argue that the children of well-off parents have always enjoyed the advantages that wealth and inheritance can confer. Is it such a leap, they ask rhetorically, to want to pass on genetic gifts to their children along with material riches? Advocates ask us to consider the positive side of germline enhancement, even if it gives an advantage to the children of those who can afford the technology. "What about . . . increasing the number of talented people? Wouldn't society be better off in the long run?" asks Dr Burke Zimmerman.

While the notion of consumer choice would appear benign, the very idea of eliminating so-called genetic defects raises the troubling question of what is meant by the term "defective." Ethicist Daniel Callahan of the Hastings Center penetrates to the core of the problem when he observes, "behind the human horror at genetic defectiveness lurks . . . an image of the perfect human being. The very language of "defect," "abnormality," "disease," and "risk" presupposes such an image, a kind of prototype of perfection."

When molecular biologists speak of mutations and genetic diseases as errors in the code, the implicit, if not explicit, assumption is that they should never have existed in the first place, that they are "bugs" or mistakes that need to be deprogrammed or corrected. The molecular biologist, in turn, becomes the computing engineer, the writer of codes, continually eliminating errors and reprogramming instructions to upgrade both the program and the performance. This is a dubious and dangerous role when we stop to consider that every human being carries a number of lethal recessive genes. Do we then come to see ourselves as "miswired" from the start, riddled with errors in our code? If that were the case, against what ideal norm of perfection are we to be measured? If every human being is made up of varying degrees of error, then we search in vain for the norm, the ideal. What makes the new language of molecular biology so subtly chilling is that it risks creating a new archetype, a flawless, errorless, perfect being to which to aspire – a new man and woman; like us, but without the warts and wrinkles, vulnerabilities and frailties, that have defined our essence from the very beginning of our existence.

Genetically Correct Politics

A spate of new scientific studies on the genetic basis of human behavior and the new sociobiology that favors nature over nurture are providing a cultural context for the widespread acceptance of the new biotechnologies. Researchers are already linking an increasing number of mental diseases to genetic disorders. Some scientists are even beginning to suggest that various form of antisocial behavior, such as shyness, misanthropy and criminality, may be examples of malfunctioning genes. The Minnesota Center for Twin and Adoption Research has found that heredity plays a determining role in a number of common personality traits. The Center has gone so far as to publish studies estimating the extent to which heredity determines personality: tendency to worry, 55 percent; creativity, 55 percent; conformity, 60 percent; aggressiveness, 48 percent; extroversion, 61 percent. Many sociobiologists go even further, contending that virtually all human activity is, in some way, determined by our genetic make-up, and that if we wish to change our situation, we must first change our genes.

The accumulating body of studies on the genetic links to personality and behavior is having an effect on public discourse. It is important to remember that from the end of the Second World War through the 1980s, social scientists argued that it is only by instituting changes in the environment that social evils could be addressed. The orthodox political wisdom has favored nurture over nature. Now, plagued by deepening social crises, the industrial nations seem no longer able to make significant changes by the traditional path of institutional and environmental reform. The sociobiologists and others of their persuasion contend that attempting to overhaul the economic and social system is at best palliative and, at worst, an exercise in futility. The key to most social and economic behavior, they maintain, is to be found at the genetic level. To change society, they therefore claim, we must first be willing to change the genes, for, while the environment is a factor, the genes are ultimately the agents most responsible for individual and group behavior.

A few lone voices in the biology community continue to caution their colleagues that they are playing fast and loose with genetics, and in the process providing grist for a new and potentially dangerous political agenda. Dr Jonathan Beckwith, a professor of microbiology and genetics at Harvard University and one of the early pioneers in the field of molecular biology, argues that a more balanced presentation of the relationship between genetics and environment needs to be made in the public arena, lest we risk the new science becoming the handmaiden for a eugenics-based politics. Beckwith points out that many diseases, such as cancer and depression, are the result of the subtle and not-so-subtle interactions of genetic predispositions and environmental triggers, and to ignore the relationships and concentrate only on the gene is tantamount to abandoning any idea of moderating or reforming the environment as a remedial strategy. The political implications are significant:

> [T]he focus on genetics alone as explanatory of disease and social problems tends to direct society's attention away from other means of dealing with such problems . . . Genetic explanations for intelligence, sex role differences, or aggression lead to an absolving of society of any responsibility for its inequities, thus providing support for those who have an interest in maintaining these inequities.[6]

Despite the fact that new experimental research is undermining the arguments and assumptions based on simple genetic reductionism, the idea of the "master molecule" that controls our biological destiny has proven so useful in advancing the interests of the molecular biologists and the many commercial firms that make up the biotech industry that it continues to gather momentum, both in the media and in public discourse, as an explanatory tool for understanding personality development, adolescent behavior, ethnic and racial differences, collective psychology, and even the workings of culture, commerce and politics.

Already, genetic information is being used by schools, employers, insurance companies and governments to determine educational tracks, employment prospects, insurance premiums and security clearances, giving rise to a new and virulent form of discrimination based on one's genetic profile. Even more chilling, some genetic engineers envision a future with a small segment of the human population engineered to "perfection" while others remain as flawed reminders of an outmoded evolutionary design. Molecular biologist Lee Silver of Princeton University writes about a not-too-distant future made up of two distinct biological classes which he refers to as the Gen Rich and Naturals. The Gen Rich, which account for 10 percent of the population, have been enhanced with synthetic genes and have become the rulers of society. They include Gen Rich businessmen, musicians, artists, intellectuals and athletes, each enhanced with specific synthetic genes to allow them to succeed in their respective fields in ways not even conceivable among those born of nature's lottery.

Today, the ultimate exercise of power is within grasp: the ability to control, at the most fundamental level, the future lives of unborn generations by engineering their biological life process in advance, making them a "partial" hostage of their own architecturally designed blueprints. I use the word "partial" because, like many others, I believe that environment is a major contributing factor in determining one's life course. It is also true, however, that one's genetic make-up plays a role in helping to shape one's destiny. Genetic engineering, then, represents the power of authorship, albeit "limited" authorship. The ability to engineer even minor changes in the physical and behavioral characteristics of future generations represents a new era in human history. Never before has such power over human life even been a possibility.

Human genetic engineering raises the very real specter of a dystopian future where the haves and have-nots are increasingly divided and separated by genetic endowment, where genetic discrimination is widely practiced, and where traditional notions of democracy and equality give way to the creation of a "genetocracy" based on one's "genetic qualifications."

A Personal Perspective

Over the past twenty-five years I have expressed growing concern about many aspects of the emerging biotech revolution, leading many in the scientific community, and in the general public, to ask if, in fact, I am simply opposed to science and the introduction of new technologies. The question is not whether one is in favor or opposed to science and technology writ large, but rather, what kind of science and technology one does favor.

We have become so accustomed to thinking of science in strictly Baconian terms that we have lost sight of other approaches to harnessing the secrets of nature. Bacon viewed nature as a "common harlot" and urged future generations to "tame," "squeeze," "mold" and "shape" her so that "man" could become her master and the undisputed sovereign of the physical world. Many of today's best-known molecular biologists are heirs to the Baconian tradition. They see the world in reductionist terms, and view their task as grand engineers, continually editing, recombining and reprogramming the genetic components of life to create more compliant, efficient and useful organisms that can be put to the service of humankind. In their research, they often favor isolation over integration, detachment over engagement, and the exercise of applied force over stewardship and nurturing.

Others in the field of biology, although equally rigorous, exercise a more integrative, systemic approach to nature. The ecological sciences, which are gaining in stature and importance, view nature as a seamless web made of myriad symbiotic relationships and mutual dependencies, all embedded in larger biotic communities that together make up a single living organism – the biosphere. Ecologists favor more subtle forms of manipulation designed to enhance rather than sever existing relationships, always with an eye toward preserving ecological diversity and maintaining community bonds.

Each of these approaches to the biological sciences leads to very different kinds of practices. For example, in agriculture, molecular biologists are experimenting with new ways to insert genes into the biological code of food crops to make them more nutritious and more resistant to herbicides, pests, bacteria and fungi. Their goal is to create a self-contained, safe haven, fortressed away from the larger biotic community. Many ecological scientists, on the other hand, are using the new flow of genomic data to understand better the relationship between environmental influences and genetic mutations to advance the science of ecology-based agriculture. Their goal is to combine the wealth of new genetic information being collected with the knowledge being gained on how ecosystems function, to establish a more integrative approach to agriculture – one that relies on diversified pest management, crop rotation, organic fertilization, and other sustainable methods designed to make agricultural production compatible with the ecosystem dynamics of the regions where the crops are being grown.

Similarly, in medicine, molecular biologists are fixing their attention on somatic gene surgery, pumping altered genes into the patient to "correct" disorders and arrest the progress of disease. Their efforts are designed to cure people who have become ill. Other researchers, however, including a small but growing number of molecular biologists, are exploring the relationship between genetic mutations and environmental triggers with the hope of fashioning a more sophisticated, scientifically based understanding and approach to preventive health. More than 70 percent of all deaths in the United States and other industrialized countries are attributable to what physicians refer to as "diseases of affluence." Heart attacks, strokes, breast, colon and prostate cancer, and diabetes are among the most common diseases of affluence. While each individual has varying genetic susceptibilities to these diseases, environmental factors, including diet and lifestyle, are major contributing elements that can trigger genetic mutations. Heavy cigarette-smoking, high levels of alcohol consumption, diets rich in animal fats, the use of pesticides and other poisonous chemicals, contaminated water and food, polluted air and sedentary living habits with little or no exercise, have been shown, in study after study, to cause genetic mutations and lead to the onset of many of these high-profile diseases.

The Human Genome Project is providing researchers with vital new information on recessive gene traits and genetic predispositions for a range of illnesses. Still, little research has been done, to date, on how genetic predispositions interact with toxic materials in the environment, the metabolizing of different foods, and lifestyle to effect genetic mutations and phenotypical expression. The new holistic approach to human medicine views the individual genome as part of an embedded organismic structure, continually interacting with and being affected by the environment in which it unfolds. The effort is geared toward using increasingly sophisticated genetic and environmental information to prevent genetic mutations from occurring. In short, the "hard-path" approach uses the new genetic science to engineer radical changes in the very blueprint of species to advance progress, while the "soft-path" approach uses the same genetic science to create a more sustainable relationship between existing species and their environments.

Each vision of science outlined here is based on different sets of human values, although I suspect that most molecular biologists continue to entertain the notion that their approach is unbiased, objective and the only true science. Notwithstanding their remonstrations, what you see ultimately depends on what you are looking for. The search is always preconditioned by the bias of particular researchers.

The splitting of the atom and the unraveling of the DNA double helix represent the two premier scientific accomplishments of the twentieth century, the first a tour de force of physics, the second of biology. Both, when applied in the form of new technologies, represent unparalleled potential power to alter both the physical and natural worlds. In the case of nuclear technology, in the form of the bomb and nuclear energy, some nations belatedly chose to reduce and even discontinue their production and use, concluding that the risk in deployment, both to the environment and to current and future generations, exceeded any potential benefits. Only two atomic bombs have been dropped on human populations in more than a half-century. Nuclear energy, once considered the greatest source of power ever developed, has been partly or largely abandoned in many countries for financial and environmental reasons.

While it might seem highly improbable, even inconceivable, to most of the principled players in this new technology revolution that genetic engineering, with all of its potential promise, might ultimately be rejected, we need to remind ourselves that, just a generation ago, it would have been equally inconceivable to imagine the partial abandonment of nuclear energy, which had for years been so enthusiastically embraced as the ultimate salvation for a society whose appetite for energy appeared nearly insatiable. It is also possible that society will accept some and reject other uses of genetic engineering in the coming biotech century. For example, one could make a solid case for genetic screening – with the appropriate safeguards in place – to better predict the onslaught of disabling diseases, especially those that can be prevented with early treatment. The new gene-splicing technologies also open the door to a new generation of lifesaving pharmaceutical products. On the other hand, the use of gene therapy to make corrective changes in the human germ line, affecting the options of future generations, is far more problematic, as is the effort to release large numbers of transgenic organisms into the Earth's biosphere. Society may well agree to some of the genetic engineering options and reject others. After all, nuclear technology has been harnessed effectively for uses other than creating energy and making bombs.

A rejection of some genetic engineering technologies does not mean that the wealth of genomic and environmental information being collected cannot be used in other ways. While the twenty-first century will be the Age of Biology, the technological application of the knowledge we gain can take a variety of forms. To believe that genetic engineering is the only way to apply our new knowledge of biology and the life sciences is limiting. It keeps us from entertaining other options that might prove even more effective in addressing the needs and fulfilling the dreams of current and future generations.

It needs to be stressed that it is not and never has been a matter of saying "yes" or "no" to the use of technology itself – although many in the scientific establishment like to frame the issue this way, leaving the impression that if one is opposed to their particular technological vision, one is anti-technology. In this sense, their position on technology mirrors their position on science, in both cases taking a fundamentalist view that there is only one "true path" to the future.

Rather, the question is what kind of biotechnologies will we choose in the coming biotech century? For example, will we use our new insights into the workings of plant and animal genomes to create genetically engineered "super crops" and transgenic animals, or new techniques for advancing organic agriculture and more humane animal husbandry practices? Will we use the information we are collecting on the human genome to alter our genetic make-up or to pursue new sophisticated practices in preventive medicine.

We may decide, in the final analysis, to shift technological priorities altogether. Now, the hard-path genetic engineering technologies are the dominant mode of application of the new biological sciences. The more integrative and embedded soft-path biotechnological applications, the ones more sensitive to ecosystem dynamics and interrelationships, remain marginal to the unfolding of the biotech century. However, it is not difficult to imagine a turnaround of sorts, in the years ahead, with the more ecological and preventive health biotechnologies taking precedence and with some genetic engineering technologies being abandoned and others used in a limited fashion and only as options of last resort. For example, in those cases where prevention and holistic health practices are insufficient to ward off seriously debilitating or deadly genetic diseases, somatic gene surgery may be an appropriate remedy.

We should also consider the very real possibility that the new genetic engineering technologies may not, in the final analysis, deliver on many of their promises. The reason for saying this is because most molecular biologists are still wed to the older, industrial frame of mind. They continue to try to force living processes into linear contexts, believing it possible to manipulate development, gene by gene, as if an organism were merely an assemblage of the individual genes that constitute it. This old-fashioned reductionist approach to biotechnology, with its emphasis on sequentiality and strict causality, is likely to meet with only limited success. The biotech century will ultimately belong to the systems thinkers, those who see biology more as "process" than "construction" and who view the gene, the organism, the ecosystem and the biosphere as an integrated "super organism," with the health of each part dependent on the health and well-being of the whole system. That is why the hard-path genetic engineers might eventually lose their dominant position to the soft-path ecologists and developmental geneticists whose thinking is more in tune with a biosphere consciousness. If that were to happen, soft-path biotechnologies might yet triumph over hard-path gene-splicing techniques in the biotech century.

The biotech revolution forces us to hold a mirror to our most deeply held views about the purpose and direction of life, making us ponder the ultimate question of the meaning of existence. This may turn out to be its greatest contribution. The rest is up to us.

Soft-Path Biotech and Networked Ways of Doing Business

We are making a great transition out of the world of physics and chemistry and into the world of biology. Dramatic new discoveries in genetics are laying the groundwork for a commercial and social revolution of epochal proportions. Up to now the debate over biotechnology has been rather narrow and sophomoric. Industry advocates talk about a coming biological renaissance while critics worry about the prospect of a "Brave New World." The media often characterizes the nature of the debate in stark terms, pitting scientific progress against an uninformed Luddite reaction. The issue, however, is more complex. The science itself is valuable. Learning about genes is helping us to better understand the workings of life. The question is how do we apply this new-found knowledge in a way that best serves the needs and aspirations of future generations?

Two broad technological visions are beginning to take shape on the eve of the biotech century. Each is based on a different set of principles and assumptions of how best to harness the new genetic science. The hard path – the dominant approach currently being used by the life sciences industry – is based on creating a second genesis in fields ranging from agriculture to human medicine. The goal is to intervene directly into the evolutionary schema, and use a range of new techniques including recombinant DNA, cloning and embryo fusion to help direct the very future of evolution itself. Hard-path biotechnologies include genetically engineered food crops, super drugs for treating illness, gene therapies and, in the future, cloned body parts and designer babies. The soft path – which is just beginning to be explored in research and development – is based on using the knowledge gleaned in the new genetic science to better steward the existing evolutionary schema. The goal is to create products and processes that are sustainable and that reaffirm the intrinsic value – as opposed to the mere utility value – of all of the life forms that constitute the earth's biosphere. Soft-path researchers are using the new discoveries in genomics to establish a more sophisticated scientific approach to improving such things as organic and sustainable agricultural practices and preventive measures in the field of medicine.

Ironically, many – if not all – of the hard-path genetic technologies, while currently touted as "cutting edge," are likely to be regarded in the not too distant future as old-fashioned and far too reductionist in nature to be commercially viable and socially acceptable, while the soft-path uses of the new science are likely to gain in popularity and become the dominant commercial expression of the new science in a range of fields by the second decade of the twenty-first century.

While hard-path biotechnologies are more compatible with the operating assumptions of a market exchange economy, soft-path biotechnologies are more appropriate venues for networked ways of doing business. As the life sciences industry makes the transition from markets to networks, it will likely increasingly redirect its research and development away from hard-path biotechnologies and toward soft-path applications and solutions.

In a market, pharmaceutical companies want to sell as many drugs as possible and provide as many medical procedures as warranted. In markets, profit is made on treating illness. Today, however, the life science industry, like other industries, is facing a global market where transaction costs are diminishing, profit margins are narrowing and production capacity is being underutilized because of insufficient global demand. Some life science companies are beginning to make the transition from markets to networks and from selling goods to providing services as a way to recapture revenue.

For example, both Eli Lilly and SmithKlineBeecham have initiated prototype disease-management programs in the United Kingdom. The companies handle a number of diseases including central nervous disorders, cancer, stroke, heart attack and diabetes. The new corporate mission is preventive: to get the patient well and keep him or her well. However, if the patient is well, he or she may be using far fewer drugs and other medical procedures. How then do Eli Lilly and SmithKline make money? The pharmaceutical companies enter into a B2B networked relationship with the insurance companies, hospital and other health providers. If Eli Lilly and SmithKline can keep their clients well, it means less medical costs for the insurance companies and other health providers. The savings are shared with the pharmaceutical companies in the form of "gain savings" agreements. BUPA, the British health insurance company, and SmithKline have broadened their disease-management network to include employers. When employees are sick their productivity on the job diminishes. When employees are healthy and fit their productivity rises. The SmithKline/BUPA network has entered into B2B gain-savings agreements with employers in which the employers share a percentage of the savings resulting from increases in worker productivity to the pharmaceutical company and the other members of the network.

The above example is illustrative of the profound difference in the way money is made in networks as opposed to markets. In markets, sellers make money by managing production and by the margins of the transaction and the volume of the units sold. In networks, partners make money in exactly the opposite way – by minimizing production, pooling risks and sharing savings. In the coming century, more profit is to be made in networks by minimizing production and sharing the savings than in markets by maximizing production and by the margins of the sales transactions.

Keeping people well in networks with the introduction of soft-path preventive medicine technologies is likely to be far more commercially lucrative than treating illness with the introduction of hard-path genetically engineered drugs, surgeries and other medical procedures. Keeping people well is a 24/7 operation requiring a full-time sustained relationship between server and client, whereas treating illness is episodic and transitory in nature and therefore requires less of an ongoing relationship. In the new world of prevention, clients pay providers to help keep them well.

The knowledge that comes out of the mapping of the human genome offers the possibility of rethinking our whole approach to medicine in the coming decades. Within the next ten to twenty years, all of the thousands of genes that comprise the human genome will have been discovered and identified. Each person will be able to have a complete genetic profile. At the same time, scientists are learning more about the relationship between genetic predispositions to specific diseases and various environmental factors that play a role in triggering genetic mutations. For example, most of the major diseases that cause serious illness and death in affluent countries – heart attacks, strokes, diabetes and cancer – occur when people with a genetic predisposition are exposed to harmful environmental stimuli.

If someone has a genetic predisposition for diabetes, cancer, stroke or heart attack, and that person smokes or drinks heavily, eats large amounts of fatty meats, leads a sedentary lifestyle or lives in a polluted community, the chances are that he or she will come down with at least one of these diseases. Conversely, we are learning in the new field of genetics that certain foods and lifestyles can help prevent certain genetic predispositions for illness from occurring. Men are now told by their physicians to eat tomato sauce because it contains genes whose proteins prevent prostate cancer. Similarly, patients with a predisposition for stomach cancer are counseled to drink green tea as a preventive measure.

In the coming years, scientists will be mapping the entire genomes of thousands of plants and microbes, making it possible to customize diets – a clustering of thousands of genes – to the individual genetic profile and predispositions of each person through every passage of his or her life from conception to birth and from birth to old age. It is likely that even a number of genetically inherited diseases might be preventable with appropriate dietary and other environmental changes during pregnancy. Lifestyles, exercise and other activities can similarly be modified to meet the very specific genetic profile of each person. This valuable knowledge about genetic predisposition and environmental triggers can be harnessed to establish a wellness regime for every individual throughout his or her lifetime. The new soft-path approach to genomics could extend the average lifespan and quality of life to well into the nineties or hundreds. The money saved in making a shift from treating illness to maintaining wellness could be shared in vast commercial health networks. The prospects for the healthcare system are nothing less than revolutionary.

This does not mean that hard-path genetic techniques ought to be abandoned altogether. It does mean that they should be treated as a last option, not a first choice. In the final analysis, there is far more money to be made in keeping people well in networks and sharing the savings than in treating people for illness in markets and making money on the diminishing transaction costs.

3

Will Life Sciences be a Driving Force of the Twenty-First Century Economy? Challenges for Arabic-Speaking States

*Juan Enriquez-Cabot and Helen Quigley**

Perhaps one of the problems in the Arabic-speaking world[1] today is that it is not conservative enough. Life sciences will become a driving force of this century's global economy. It is an area where Arabic speakers have excelled in previous eras, but not in recent centuries. One of the key reasons why Arabic-speaking nations are no longer leaders in the life sciences is that they largely ignore the lessons and activities carried out within the region one millennium ago. Many of today's leaders are focused on the traditions prevalent five hundred years ago. However, the right period to look at, and emulate, is one millennium ago.

Given today's conflicts, poverty and upheaval, it is easy to forget that, for centuries, "the world of Islam was in the forefront of human civilization and achievement . . . there was only one civilization that was comparable in the level, quality and variety of achievement . . . China."[2] Arabic speakers dominated commerce, research and learning. They excelled at mathematics, biology, chemistry and astronomy (see Figure 3.1).

* The authors would like to thank Bruce Scott, George Lodge, Rodrigo Martinez and German Gaytan for their help and comments.

Figure 3.1
Men of Science

Source: Buzurg, Ibn Shahriyar, *Kitab 'ajayibal-Hind* (Leiden, Netherlands: Brill, 1886).

To launch a significant life sciences effort today, one has to bring together the leading edge of many disciplines, including biology, chemistry, mathematics, medicine and pharmacology. This is something that has already occurred, and has since been lost, in many parts of the Arabic-speaking world.

Among some people, the current stereotype is that Arabic speakers are not very talented in mathematics and science, also that they are not good and patient researchers and that only "Asians" are really good at some of these endeavors. This stereotype is the result of ignorance, of tradition and of lost education, either forgotten or ignored. It is not the result of any inherent ability or disability. Accusations of inherent ineptitude have been launched time and again against every type of people, race and region for centuries. After all, in the 1960s many in the United States believed that "Made in Japan" was a synonym for bad quality . . . and that things would always remain this way. In the 1980s few would have associated the words "high tech" with "Ireland."

Countries, regions and companies change over time. This happens with increased frequency as wealth is generated not by inherited assets but through new knowledge. Whether or not the Arabic-speaking peoples play a leading role in the life sciences revolution depends on decisions they make for themselves and for their children, not on inherent talents that they have or do not have.

In terms of overall competitiveness, it is worth examining where various Islamic and non-Islamic countries have stood in the past and where they stand today. Within this context, it is possible to review whether life sciences might play a role in the overall development strategy of some Middle Eastern states. More specifically, the question of whether this area of research should become a part of regional development strategy can then be addressed. There are hints throughout history of what some Middle Eastern states might have accomplished had they chosen to do so. More recently, several small states throughout various regions, including some within the Middle East, have become very successful.

However, let us first, briefly, review the ways in which people and states have generated wealth. It is worthwhile for current leaders to understand the important role played initially by agriculture and manufacturing, currently by the digital revolution and increasingly by life sciences.

Wealth Generation in Today's World

We often think of states as permanent and powerful, but in the context of human history, they are relatively new and for the most part quite fragile. Through 1500, only about one-fifth of the world's land mass was divided up into states "run by bureaucrats and governed by laws."[3] Even today, though the whole of the globe has been carved into states, the overall situation is far from stable.

Over the past centuries, and particularly over the past eight decades, many great empires, civilizations and nations collapsed or split apart despite powerful military machines and relatively good governments. This occurred across Africa, Asia and Europe. It happened to rich and to poor states. It occurred in states with fundamental religious and ethnic divides, and also in those that thought they addressed these cleavages long ago. One might expect secessionist debates to rage across a broad archipelago like Indonesia but it is jarring to see similar debates taking place within places like the British Isles, Belgium, Italy, Canada and Spain.[4] No state is immune.

One key reason why there are almost four times as many countries in the world today as there were fifty years ago is because most governments fail to understand fundamental and ever-faster changes in the way wealth is generated. This basic failure to understand the new rules of economics leads to a proliferation of sovereign and unequal desks within the United Nations General Assembly. After all, the body was founded with fifty members and now numbers one hundred and ninety-two.

One should not assume that what exists today will also remain tomorrow. It is no longer necessary to be part of a large mass to be economically powerful. In fact, some of the world's richest people live in some of its smallest and most resource-poor states. Citizens of Luxembourg can produce almost 30 percent more wealth each year than those of the United States. Meanwhile, some states that are rich in natural resources are becoming much poorer because they are not educating their youth in ways that allow the generation of new knowledge and therefore new wealth (see Figure 3.2).

Figure 3.2
GDP Per Capita: Saudi Arabia (US$)

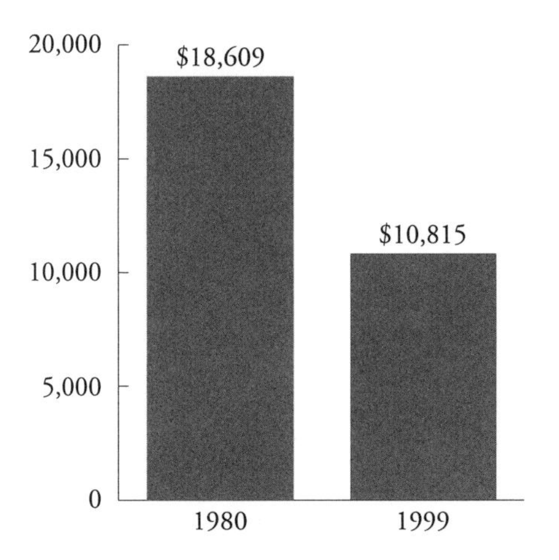

Source: Human Development Report, 2001.

States and wealth are quite recent creations. For most of the four billion odd years that life has been on the planet, understanding how to generate, keep and accumulate vast amounts of wealth was a mostly irrelevant notion. Humanoids spent their first five to six million years desperately hunting and gathering enough to meet their basic needs rather than investing in the future by opening bank accounts and investing in stocks.

This nomadic foraging was not the best way to use our brains. As a species we came very close to becoming extinct. At one point, some scientists estimate that there were fewer than twelve thousand of our species left wandering the planet.[5] Various parallel lines of hominoids, some of which coexisted with our own during some periods, did go extinct.[6] In this world the "rich person" gradually became the one who could find and catch an animal or gather fruits and fiber most effectively. The fleeter and the stronger tended to dominate protein production.

Roughly ten thousand years ago, food production and therefore wealth creation started becoming more deliberate. Villages emerged. Understanding and applying ever-greater amounts of knowledge became as important as speed and brawn. Very gradually, societies began to accumulate excess food, which meant they could store wealth and dedicate part of their time and resources to building even more wealth as well as to building cities and ceremonial centers. However, it was hard for any given individual to produce that much wealth overall, so as empires formed and grew, the name of the game was to have more children, control more natural resources, accumulate more land. For hundreds of years, the world's great powers were countries with large populations like India and China and vast empires like that of Rome. Just before the Industrial Revolution, India and China alone accounted for nearly 40 percent of world economic output.[7]

As states fought to take over existing resources, there was little economic growth overall. Individual rulers and countries could accumulate a lot of wealth, but global production and wealth remained static. Between 1000 and about 1820, global economic growth per capita averaged about 0.1 percent per year.[8]

A few thousand years ago, the areas best able to generate capital and accumulate wealth did so through new life sciences technologies. It was the ability to breed and produce ever more productive livestock and crops that gave rise to some of the world's great early writing, art and science. This revolution first occurred on a large scale within the Middle

East, in an area known as "the fertile crescent." A key component of this economic and cultural dominance was life sciences, since this was the first area to domesticate livestock such as goats, sheep and cows.[9] What began as a serendipitous economy for people living where much of the world's most productive plant and animal species lived, eventually turned into a vast civilization based on increasing knowledge of astronomy, trade, hydrology, weapons and construction.

However, learning that worked once upon a time is eventually superseded by further discovery. Societies that do not remain open, quit learning, and continue to depend on what they were given rather than on what they continuously generate, tend to lose prominence, and eventually collapse. Many countries forget these lessons, even though they are exemplified and reiterated time and again by the collapse of empires. Such countries assume that land and "commodities," rather than brains and new ideas, power progress. They fight bloody wars to prevent any secession or change, to protect mines and minerals, and to expand their borders. They promote rapid population growth regardless of whether they have the educational and healthcare institutions to support such growth. Consequently, they become poorer every year, despite having what used to be considered very valuable natural resources on a massive scale.

Some never understand how the rules have changed and what it means for them. True, the greatest fortunes in the United States used to come from oil. Few approached John D. Rockefeller's power and wealth when he headed Standard Oil. However, as the means of generating great wealth changed, so too did the nature of the most powerful companies and individuals.

It is no longer oil that backs the greatest US fortunes, nor that forms the backbone of the globe's most powerful countries. Neither do other natural resources. A company like Microsoft or IBM owns little land, virtually no natural resources, and employs relatively few people per dollar of wealth generated. At its peak, in recent decades, oil only represented about 4 percent of the world's wealth. Today, it represents less than 1.5 percent (see Figures 3.3 and 3.4).

How Important is Oil Overall?[10]

Figure 3.3
Worldwide Oil Production (1,000 barrels per day)

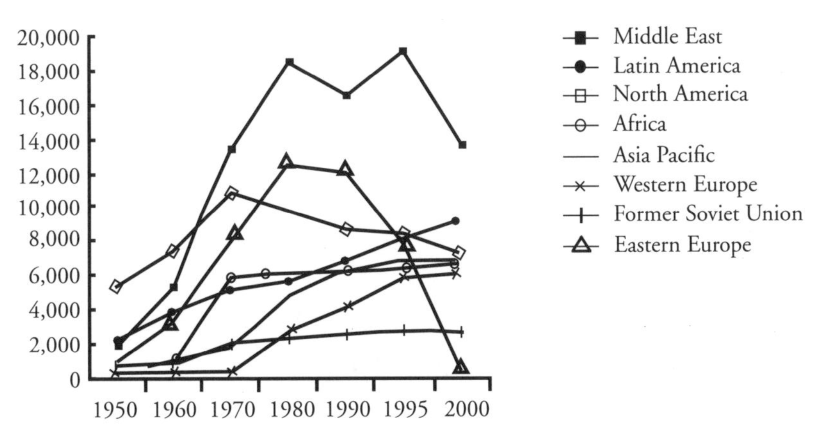

Source: *International Petroleum Encyclopedia*, 2002, pages 211–215.

Figure 3.4
Oil Production as a Percentage of World Gross Production

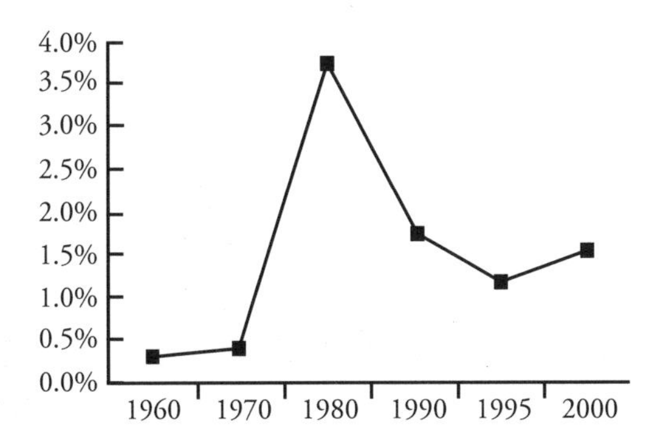

Note: The price used for this calculation was the annual average wellhead price of united crude oil.

Source: World Bank, *World Development Indicators, 2002, CD ROM. International Petroleum Encyclopedia, 2002. Basic Petroleum Data Book, Petroleum Industry Statistics.* Volume XXII, Number 2, August 2002.

While many countries failed to understand these trends in wealth creation, a few did begin to change quickly.

From Digital Dominance Towards a Genomic Future

What substituted for "commodities" was bits,[11] or in other words, ideas. These also required building a different set of institutions, changing notions of human rights, building complex and partly open trade regimes, and an increasing emphasis on the creation and protection of intellectual property.

In this new intellectual property (IP) regime, the dominant language is neither English nor Chinese, neither Russian nor Arabic. It is 1s and 0s. This alphabet was not so important in 1970. Today, it accounts for 93 percent of all data transmitted on the planet.[12] It was the countries that learned and applied this new language and educated their population in how to exploit this new resource that became rich.[13] One can observe this effect in the income per capita trends given below, even when compared using purchasing power parity (PPP) (see Table 3.1).

Table 3.1
GDP Per Capita for Selected Countries
(Purchasing Power Parity)

	GDP p/c PPP, Selected Countries (in US dollars)					
	1975	*1980*	*1985*	*1990*	*1995*	*2000*
Ireland	3,435	5,841	8,051	12,687	17,921	29,866
Singapore	2,847	5,917	7,662	12,783	19,432	23,356
Israel	5,270	7,908	9,857	13,450	17,550	20,131
Korea, Rep.	1,693	3,037	4,906	8,880	13,814	17,380
Islamic Avg (34)	1,371	2,143	2,515	3,457	4,553	5,511

Source: World Bank, *World Development Indicators 2002* (CD ROM).

Some believe that the digital era is over and that it generated little wealth. The dot-com bust notwithstanding, few of us have stopped using e-mail, researching online or buying off the Internet. The alphabet we use to transmit our phone conversations, music, pictures, films and medical records has changed. So too have the countries we consider present and future leaders. Remaining digitally illiterate is ever more expensive for a society.

In 1965, few would have bet on Japan, Korea, Singapore or Taiwan as examples to follow. The World Bank, for one, was betting on Burma (today Myanmar) and the Philippines. The first group of states learned a new alphabet and began producing robots, cars, computers, ships, financial and insurance products. The second group continued to depend on their vast natural resource wealth. Their relative position in the global economy changed very quickly. For the past few decades, most new wealth has come from managing data, not machines. Even grand manufacturing-based companies, like General Electric, make most of their profits from areas like GE Capital rather than giant turbines. In fact, my colleague Bruce Scott has researched and written extensively on these trends.[14]

In a knowledge economy, it is not necessary to be large, nor be historically successful. What matters is the ability to be smart, flexible, open, and be able to educate and/or attract entrepreneurs who understand the new, dominant alphabets. It must also be recognized that when the dominant alphabet begins to change yet again, the institutional frameworks will have to change, which is precisely what is occurring right now.

While 1s and 0s have been the engine of growth and have allowed deserts like Silicon Valley and Bangalore to bloom, now we are beginning to understand how to read the four letters that code for all life. In April 1953, five decades ago, Watson and Crick discovered the structure of the molecule that codes every bacterium, plant, animal and human on the planet, DNA.[15] This molecule is a long series of four base pairs, adenine, thiamine, cytosine and guanine (A,T,C,G) supported by a sugar phosphate backbone.

This is a far more powerful alphabet than 1s and 0s. It allows us to start directly and deliberately recoding life forms, and when we recode in certain ways, the function executed by a life form changes. Just as the digital revolution changed the relative position of nations, so too will the genomics revolution.

The Emerging Biotechonomy[16]

Our expanding knowledge of how the genetic alphabet executes various functions is likely to change the rules in most basic industries. Just as a diskette or CD changes what it does based on the order of a string of

1s and 0s, a seed will change what it grows and produces based on the order of its A,T,C,Gs.

One can already reprogram a goat's genetic code (genome) so that it expresses spider's silk protein within its milk.[17] This in turn, as we shall see, changes the nature and value of agriculture, medicine, fishing and war. Who is likely to become richer: a traditional goat farmer or one whose goats produce a valuable textile material within its body? The latter farmers live in a few towns in Massachusetts and Canada today. Each of their goats is treated royally. Each animal is worth thousands of dollars. The value is derived from the varied uses of its new product.

A few changes even in a single species can have a broad impact across industries. When you harvest spider's silk from goats, doctors begin to pay attention to agribusiness, among other reasons because they can begin to access and use ultra-fine, very strong, biodegradable sutures during surgery. (And they may begin administering vaccines grown in bananas, potatoes and tobacco, or they may prescribe medicines grown within a goat's body for that matter.)

Environmentalists also begin to pay attention to goats because now they can ask fishermen to stop using plastic and begin using biodegradable fishing lines. Generals are also intrigued. Their soldiers may begin to wear shirts spun from spiders' silk. This makes them ultra-light, flexible, and perhaps four times as strong as current bullet-proof vests.

Just as 1s and 0s changed the world, so too will A,T,C,Gs. Our knowledge of this alphabet is increasing daily at a rate that makes the digital revolution seem slow.

Great libraries have always reflected great power and knowledge. Ancient Alexandria was a global center of learning, as were Damascus and Baghdad (see Figure 3.5). Thousands of volumes and scholars reflected accumulated wisdom, travels, conquests and, above all, the ability to adopt, adapt and transmit useful knowledge to future generations. However, many of the great libraries and centers of global learning have been burned, abandoned, sacked, or have moved away from many Arabic-speaking states. Many of the great early science texts are abroad. Yet this does not mean that people in the region do not have access to great libraries and data.

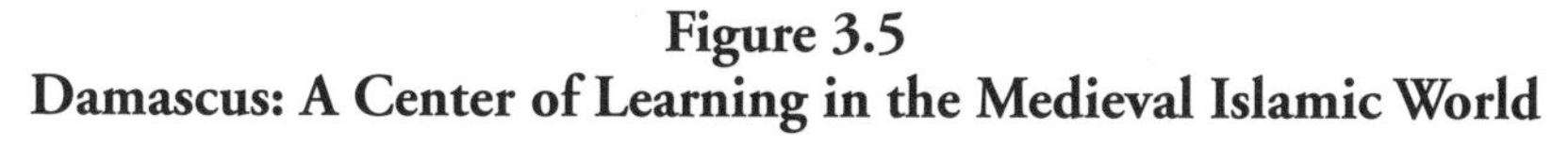

Figure 3.5
Damascus: A Center of Learning in the Medieval Islamic World

Source: J.S. Buckingham. *Travels Among the Arab Tribes Inhabiting the Countries East of Syria and Palestine*, London, 1825.

Today's greatest libraries are not based on the ABCs. Most data is stored in 1s and 0s. Some of the biggest libraries are non-territorial and geographically unbounded. A resource like Google accesses more data, faster, than any paper-based library anywhere.[18] In February 2003 it claimed to search 3.083 billion pages in seconds. This means that countries and regions do not have to make a huge physical investment in collecting processed paper pulp covered with symbols. They can access and use enormous depositories of knowledge as long as they invest and motivate their own citizens and attract smart foreigners.

Knowing how to access and apply vast databases becomes even more crucial as existing databases of words, images, and sounds are swamped by waves of biodata. As we learn the basics of a new, more universal alphabet – that of all life forms – we are generating a great deal of new data. Today's bio-libraries are based on data-like gene code, proteins and single nucleotide polymorphisms. Already, bio-libraries hold much more data than the largest printed depository of knowledge in the world: the physical collections of the Library of Congress in the United States. Tomorrow's bio-libraries will be far larger. Computer power will have to increase massively to accommodate an increasing number of still and moving pictures.

Today, a single genomics company can generate data equivalent to several times the total of printed materials in the Library of Congress. And this data compounds every month. As the information in these data banks is annotated and correlated, it compounds in size and complexity. It used to be considered good work for a top chemist to produce a few new chemical compounds per year. Today a smart graduate student at the Harvard Medical School produces over seven thousand new compounds in six months and analyses them across seven hundred and twenty dimensions.[19]

About half of this new mass of genetic sequence data is available for free.[20] It is in the public domain and can be accessed from your own computer. You can go back to your office and begin to download billions of bits worth of gene data from Japan, Europe or the United States.[21] This will tell you how diseases like HIV, the flu and meningitis are coded. It will allow you to begin comparing plants, fish, cows and chimps with humans.

Understanding our own genome, and more complex structures like our proteome (identification of proteins and their interactions), may help us figure out who gets sick and who does not, who gets cured and who does not.

The impact of genomics and proteomics will go far beyond the doctor's office and pharmaceuticals. Life code is already flowing into computers in insurance companies to help understand and update actuarial tables. This is generating a healthy debate among patient advocate groups and legislators and will likely lead to significant reforms in privacy statutes.

Agribusiness is forming various partnerships with chemical, pharmaceutical, cosmetic, energy and information-management companies. Given the increasing emphasis on manufacturing through organisms, how we make things is likely to change just as much as it did during the Industrial Revolution.[22]

The data is there and change is occurring. As yet, very few understand what it means or why it is important. However, some are learning fast. These are the types of historic upheavals in the mode of production that should not be ignored. The Industrial Revolution upended India and China's global economic dominance and allowed the rise of Britain and Western Europe. The digital revolution allowed a few East Asian states to rapidly achieve developed status. Only a few leaders have the vision to change their overall institutional framework, laws, education, infrastructure and investment policies. Yet, those who do so can build great countries in a very short period of time. The genomic revolution will help determine who is dominant in areas like chemicals, pharmaceuticals, energy and agribusiness.[23]

To convert to a digital mode, one had to learn a new language and skill set. Otherwise it would be very hard to use a computer or communicate by e-mail. Today, one must learn a new set of skills to go genomic. While it may not be necessary to become a bench scientist, nor a life programmer, one must understand why the life code is important, what you can build with it and what changes it brings.

Can and should Arabic-speaking peoples make a significant effort in this area? Is there any evidence that they could succeed?

The Roots of Life Sciences in the Islamic World

Approximately one thousand years ago, coinciding with the rise of Islam, the Arabic-speaking world embarked upon a period of unsurpassed scientific discovery and achievement. The activities of this period, between the ninth and the fourteenth centuries, produced many great works in a number of disciplines.

During this golden era, Islamic scientific culture flourished, as did learning and scholarship. Advances in mathematics helped push astronomy and advances in chemistry helped medicine. The basic disciplines that are recognized today as important to the development of life sciences, including botany and anatomy, flourished. In fact, today's Western science rests partly on these discoveries and refinements.

In the medieval Islamic world, two of the greatest centers of learning and culture were Baghdad and Damascus. Other notable centers of learning were Toledo, Cordova and Ghazni. It is not impossible to change course tomorrow and rekindle the learning and knowledge that once made this region the world's center of knowledge. It has happened in other countries. It is happening today, in places like China and India.

As a case in point, during the 1970s, Vietnam was not the world's hot spot for tourism. Around thirty years later, in 2003, the following headline appeared in the *New York Times*: "Vietnam Becomes a Safe Haven for Tourists."[24] Things can and do change, sometimes for the better, and they do so ever faster.

The Islamic world has a long-standing tradition of learning and scholarship. By the ninth century, the Abbasid caliphate in Baghdad had expanded paper production and established uniform notation systems. This allowed more people to become literate and to spread ideas. Book production increased substantially. Greek and Sanskrit science texts were translated and spread. Caliphs and Sultans became patrons of art and science. Libraries flourished. In 830, Abbasid caliph Al-Ma'mun founded the Bayt Al-Hikma (the House of Wisdom) in Baghdad, which served as a library and center of learning as well as an active school and astronomical observatory.

As culture shifted from the oral towards the written, it became easier to gather, study and spread knowledge. By the tenth century, Abbasid vizier Ibn Muqla had developed the rounded Arabic script (*naskh*), the precursor of modern Arabic typography. A standard script made it even easier to print and teach. In the tenth century, Dar Al-Ilm (the House of Knowledge) was said to house over ten thousand volumes of science texts, making it one of the world's most important depositories of learning.

Unfortunately, this period also demonstrated that a knowledge center could be fragile. Much of Dar Al-Ilm's collection was lost when the Seljuks burned the library in 1055. Nevertheless, various great libraries continued to be founded and flourished throughout the Islamic world. At a time when Oxford and Cambridge were small farm towns, Persian vizier Nizam Al-Mulk founded the Nizamiyah in 1065, which still remains a model for modern institutions of learning throughout the Islamic world today.

Cairo, Mosul, Basra and Shiraz proudly hosted some of the greatest libraries of the medieval period. Knowledge also spread broadly beyond the region. Within Europe, during a period of darkness, plague and some fundamentalism, the beacon of learning was the great library of Cordoba. What little science spread throughout the West during this period became available in a very roundabout fashion. Jewish scholars worked in Spain to translate Arabic texts into Hebrew. A number of these texts would reach the few interested Western scholars who would then translate the Hebrew into Latin. Most ended up stored within monasteries far from a predominantly illiterate population. Recent popular books and exhibits like Dick Teresi's *Lost Discoveries*, Seyyed Hossein Nasr's *Islamic Science*, and the UNESCO show on Islamic Science have detailed how much was discovered and then forgotten.[25] It would be illuminating to reiterate a few of the examples they provide.

Along with the rise of Islam came the building of mosques that served as centers where religion and scientific learning successfully coexisted. The disciplines that were pursued under their domes enabled life sciences research. By the eighth century, the first hospital-school of medicine had opened in Baghdad. This was quickly followed by other hospitals in Damascus, Aleppo, Cairo and Marrakech. While much of this early medicine was initially modeled on human biology as understood by the Greeks, particularly Aristotle and Galen, the base of knowledge and research soon expanded and resulted in treatises on subjects like hygiene, plague, scurvy, skin diseases, colic, hemorrhoids, ophthalmology and fever. Many of these treatments became models for early European medicine.[26]

For centuries, medical science flourished in the Islamic world. Medical dictionaries and therapeutic manuals abounded. The head of Baghdad's hospital, Al-Razi (sometimes referred to by his Latinized name Rhazes), compiled an encyclopedia, *Al-Ḥawi*, which detailed what was known, through the tenth century, about medicine. In addition, Al-Razi was perhaps the first to observe and describe the difference between smallpox and measles.

Better known throughout the West was Avicenna, Ibn Sina, whose tenth-century *Canon on Medicine* (*Al-Qanun fi'l-tibb*), was used for centuries as a comprehensive medical text. These general compilations were then complemented by further research on specific parts of the body. Ibn Al-Nafis (thirteenth century) detailed the anatomy of the heart and the circulatory system during the thirteenth century. A century later, Ibn Ilyas drew and detailed some extraordinary descriptions of the bones of the lower jaw, diagrams of cranial sutures, and nerves. Some of the original drawings can be viewed in Washington, DC's National Library of Medicine (see Figure 3.6).[27]

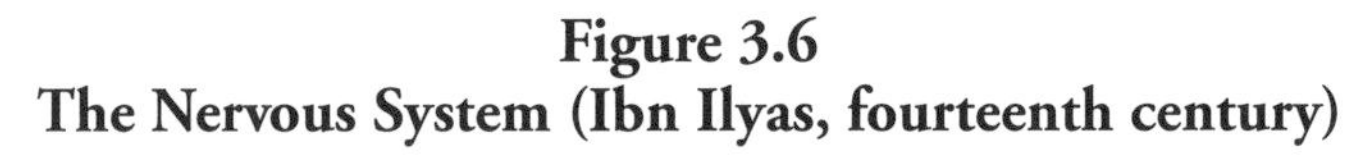

Figure 3.6
The Nervous System (Ibn Ilyas, fourteenth century)

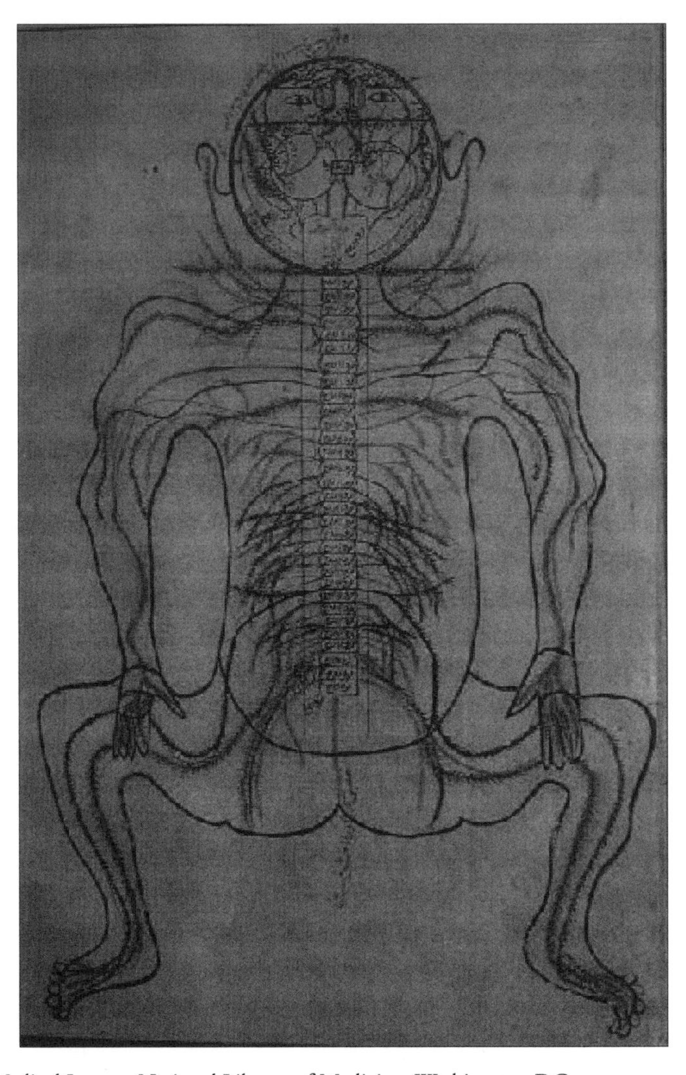

Source: Medical Images, National Library of Medicine, Washington, DC.

Hand in hand with medicine was research in botany. Here too, we find extraordinary advances. In tenth-century Jerusalem, physician-botanist Al-Tamimi produced complex drugs and published treatises on various medicinal compounds. Among the various drugs mentioned one can find a formula for the production of theriac, an anti-toxin made from venom of snakes and scorpions. There are further examples of early pharmaceutical-botanical knowledge: Al-Tabari wrote *Firdaws Al-Hikma* (*The Paradise of Wisdom*), a text on the medicinal uses of plants; Ibn Al-Baytar compiled a comprehensive encyclopedia of botany; Al-Dinawari penned the *Kitab Al-Nabat* (*Book of Plants*), and during the thirteenth century Al-Malaqi wrote treatises on botany and pharmaceutics.

Some used their botanical-scientific knowledge to advance agriculture. Ray Goldberg, who coined the term agribusiness and is one of the world's leading authorities on advanced agricultural production and integration, would have been impressed, had he lived in the eighth century, by the likes of Jabir Ibn Hayyan. This sage's treatises on botany and agriculture describe plants, their growing conditions, and diverse uses.

Botany was complemented by chemistry. Research ranged chemical medicine to metallurgical studies. Once again, Jabir Ibn Hayyan contributed to the advances made in the field. He produced works detailing experiments in crystallization, distillation, acids and many other chemical processes. As scholars searched for exotic mineral deposits, they also began to research geology. This led to treatises on fossils, and the formation of mountains and of sedimentary rocks. Ibn Sina wrote on the classification of metals and minerals and Al-Biruni studied gems and drugs as well as the weights and properties of minerals. Of particular interest, for obvious reasons, were underground water and deposit systems. In a more practical vein, by 1100 one text described seven different forms of gunpowder.[28]

Even sciences that today appear to have little direct bearing on life sciences, like astronomy, were crucial stepping-stones in both a practical and a theoretical sense. The need to ascertain the direction of the holy city of Mecca from any location for daily prayers led to the study of celestial navigation. Furthermore, one had to determine the correct time of day to maintain a daily five-prayer cycle and to observe various religious festivals.

Three observatories were responsible for the core of research and established Muslim supremacy in the field of astronomy: Caliph Al-Ma'mun's observatory (ninth century, Baghdad), the Mongol observatory at Maragha, (thirteenth century, Iran) and Ulugh-Beg's observatory in Samarkand (fifteenth century, Uzbekistan). These and many other centers brought together celestial researchers, historians, translators and mathematicians. Greek classics were translated, verified or questioned, along with Indian and Iranian texts. Further research made this region a world leader in cosmology.

Scholars like Ibn Al-Haytham (Latinized as Alhazen), furthered knowledge by questioning Ptolemy's planetary theories. During the tenth century, Ibn Al-Haytham produced monumental works on optics, light refraction and reflection, the camera obscura, and atmospheric observations. His *Optical Thesaurus* (*Kitab Al-Manaẓir*), detailed various light experiments. The oldest surviving treatise with illustrations detailing the positions of fixed stars was written in 1009 by Abd Al-Rahman Al-Sufi. Nasir Al-Din Al-Tusi, a scholar at the Maraghah observatory, devised a new mathematical model of planetary motion, which in turn was further strengthened by his pupil Qutb Al-Din Al-Shirazi (who then went on to model the motion of the planet Mercury and applied Ibn Al-Haytham's theory of optics to explain rainbows).

Many of the theories and models of this period reappeared two centuries later in the works of Copernicus, and they preceded Kepler's work by centuries.These discoveries and observations included: lunar and solar eclipses, transits of sun and fixed stars, planetary positions and theories of rotation, refinement of instruments like the astrolabe, quadrant and sextant, theories on the orbit of the moon, and various techniques to measure the earth's circumference.

All of this research was partly enabled and supported by extraordinary mathematical talent and creativity. The Abbasid caliphate translated all relevant Greek texts on mathematics by Euclid, Archimedes, Apollonius of Perga and Ptolemy as well as works in Sanskrit and middle Persian. But it is perhaps in this area of the sciences where Muslim scholarship furthered overall global knowledge the farthest.

Muslims refined and applied Indian numbers and notions like '0', the decimal system, and negative numbers. These concepts were later absorbed by the West and are evident to date (i.e. Arabic numbers). Al-Uqlidisi refined methods of mathematical calculation that eliminated the need to use the finger computation and 'dust-board' methods. Then, as paper use became ever more common, mathematical equations began to be written down, studied, and refined or refuted. Extraordinary feats of calculation and deduction carried out in a few people's minds or on a temporal surface began appearing in texts that could be studied, reproduced, questioned and advanced.

Pure mathematics grew and gave birth to scholarly fields like algebra and trigonometry. Al-Khwarazmi (ninth century, Baghdad) was among the first to develop algebraic models and refine the Arabic-Hindi system of numerals. Nasir Al-Din Al-Tusi derived fundamental theorems for trigonometry. Al-Biruni wrote definitive equations on spherical trigonometry. Persia's Omar Khayyam was an accomplished mathematician and carried out research based on Euclid's parallel line theorem. Treatises flowed on number theory, fractions, computation methods and geometry. The legacy of this period of extraordinary achievement was profound and is evident through English words like 'algebra', 'cipher' and 'algorithm.'

The bottom line is that every discipline needed to fuel a life science research effort has existed, at a world-class level, within this region. Some of the world's leading thinkers in biology, medicine, chemistry, mathematics and other basic sciences were based here once upon a time. So it is fair to ask: where did all these talents go . . . and might they someday return?

Modern Challenges to Regional Development

If there was one theme to the golden era of Arabic science, it was openness. Great research flourished because this region, unlike Europe, was willing to learn from the Greeks, Mesopotamians, Egyptians and Indians. A great deal of knowledge flowed in: texts, stories, maps and people. Openness was critical. Immigrants, both Christians and Jews, communicated some of the great discoveries and research. They were tolerated and supported at a time when many were fleeing highly repressive and dogmatic governments as well as intolerant religious strictures.

For centuries, great centers of learning and research bloomed here. Libraries and observatories received vast funding and became the centerpieces of public squares. Cultural and intellectual debates flourished in some cities. However, after the shock of the Mongol invasions in the thirteenth century, the increasing military challenges of the West, and the ossification of certain forms of government and religious practice, it became harder and harder to defend learning.

As much of the Arabic-speaking region became ever more wary of outsiders, cities and states turned inward and stopped absorbing knowledge. University systems and libraries collapsed or became very conservative.

Governments became more repressive and restrictive. Scientific inquiry became a marginal concern. Basic education foundered.

Much of the Renaissance was missed or ignored. As Western learning, science and armies came together, few people within the Middle East understood the consequences of these rapid advances. As steel, cannon, shot and military strategy advanced rapidly, the sovereignty of those who did not understand and learn eroded.[29]

Some Muslim leaders tried to redress this imbalance of power by attempting to selectively gather knowledge. They sent missions to understand the military components of Western dominance. However, their own extraordinary history of scientific advance should have been enough to warn them that overall progress requires not importing a few recipes but creating an overall context for inquiry and development. One cannot think broadly and openly about mathematics without doing the same for biology and physics. So while some knowledge of new gunpowder and military formations slowly flowed into the region,[30] the first Arabic translation of Darwin's *The Origin of Species* was not published until after the First World War.

According to Lewis, memorization of the established often substitutes for thought, experiment and analysis. There were a few attempts to deal with these problems. The Indian educationist Syed Ahmad Khan founded the Aligarh Muslim University in 1875 to try to facilitate greater levels of scientific learning among Muslims. Yet, many of these efforts were isolated and temporal. To date, throughout many parts of the Muslim world, universities are more oriented towards teaching rather than research. In addition, the focus of research is inclined to be theoretical rather than applied. National support for research is historically very low (0.02 percent to 0.5 percent of GNP).[31] Most funding tends to be public, so there is little incentive to network and seek support from the private sector. Within the scientific community, intra-Arab relations are not particularly strong and there is limited global connectivity.

Of late, the world has changed extraordinarily fast. From 1820 through 1992 "population increased five-fold, per capita product eight-fold, world GDP forty-fold, and world trade 540-fold."[32] However, not everyone participated in this unprecedented growth. In fact more and more were left further and further behind. Within the Gulf Co-operation Council (GCC) countries – Bahrain, Kuwait, Oman, Qatar, Saudi and the United Arab Emirates (UAE) – more people are trying to live off existing natural resources. Despite its strategic importance, the region's relevance to the overall global economy fades to the extent that oil continues to be the primary source of wealth. In 1980, the GCC represented 8.5 percent of world trade but in 1990 it was 2.5 percent, in 2000 only 2.6 percent.[33] How can this be?

In an agricultural society, it took a long time to generate and accumulate more wealth. Therefore the income gap between the richest and the poorest societies was about five to one. Manufacturing changed the rules of the game. If one person had twelve kids and another had a thousand horsepower, it is easy to guess who would have the competitive edge. Thus a few began producing a lot, while others continued to do what they had always been doing.

Oil tragically compounded this trend. It represents a fading opportunity. Over the past few decades, much of the developed world has been focused on understanding and applying a new alphabet, that of 1s and 0s. Companies and countries that applied it effectively became very wealthy. A small entity like Nokia, which used to produce plastic boots on the very periphery of Europe retooled into the digital world and is now one of the region's most valuable assets.

Meanwhile, oil producers continued worrying and focusing on territorial security, wells, production levels and other "stuff." The best scientific brains were used to carry out engineering calculations for refineries, to analyze geological formations and to run parts of government ministries. So if one looks at overall productivity per worker from 1995–9 (see Table 3.2), it does not look that different between South Korea, Singapore, Argentina, Israel, Iraq and Ireland.[34]

However the societal outcome in each case was very different. Some societies invested far more in schools to make their citizens ever more productive and science-literate. Others did not. Bruce Scott has detailed these trends.[36] Overall quality of life outcomes are also very different in states that tended to depend on what they had rather than what they learned. Instead of educating a broad cross-segment of their population, many resource-rich states used a few workers to make the elite quite rich. However, in a broad-based, networked, knowledge economy, this is not the way most wealth is generated. Most of the world's great oil producers are not exactly examples to follow: Nigeria, Iran, Iraq, Venezuela or even the greatest of them all, the erstwhile Soviet Union.

Table 3.2
Value Added Per Worker and Educational Investments

Country	*Value added per worker per year (1980–4)*	*Value added (US$) per worker per year (1995–9)*	*Education expenditure as % GDP per capita 1999*	*Average years of schooling: males*[35]	*Average years of schooling: females*
Iraq	13,599	34,316	–	4.6	3.3
Israel	23,459	35,526	29.7	9.3	9.4
Singapore	16,442	40,674	–	7.5	6.6
Ireland	13,599	34,316	17.4	9.3	9.4
S. Korea	11,617	40,916	–	11.7	10.0
Argentina	33,694	37,480	14.7	8.8	8.9

Source: The World Bank, *2002 Development Indicators*, Tables 2.5 and 2.13.

What happened to those who began to develop a knowledge economy versus those who continued to depend on natural resources? In 1950, the average Iraqi and Argentinian produced almost 50 percent more than someone from Ireland, about twice as much as an Israeli and five times as much as a South Korean. Yet, the latter two countries have no natural resources to focus on and fight over. The only thing they could invest in was brains and technology. By 1990, Iraq had collapsed (as did Argentina by 2002), Ireland and Israel were doing well, despite decades of upheaval, and South Korea was about to take off (see Table 3.3).

Table 3.3
Rates of Growth: Some Examples of Resource-Rich vs Resource-Poor Countries[37]

Country	*GDP 1950 million US$*	*Per Capita 1950*	*GDP 1990 million US$*	*Per Capita 1990*
UAE	996	10,594	22,473	14,134
Iraq	5,407	5.171	34,238	1,882
Argentina	85,524	4,987	212,518	6,581
Ireland	10,444	3,518	38,963	11,123
Israel	3,107	2,452	47,047	10,096
Saudi Arabia	9,061	2,190	151,996	10,222
South Korea	16,622	876	384,174	8,977

Source: Based on data from Angus Maddison, *Monitoring the World Economy 1820–1992* (Paris: OECD, 2000). Table F-4, D-1b, C16.

So what is next? We can look at things like science publications and patent trends as one, out of many, possible indicators of what might come to pass. Patents and intellectual property are a necessary fuel for any knowledge-driven economy, but are not a guarantee of success. One also has to be able to implement. Research without business is not sustainable in the long term. Countries that do not make at least some of their scientists rich have fewer role models to attract the next generation (see Table 3.4).

Table 3.4
Who is Competitive in a Knowledge-Driven Economy?

Country	*Growth in output (1980–90)*	*Growth in output (1990–2000)*	*Science journal articles (1997)*	*Patent applications filed by non-residents (1999)*[38]	*Patent applications filed by residents (1999)*
Iran	1.7	3.5	332	177	366
Iraq	-6.8	–	35	–	–
Israel	3.5	5.1	5,321	46,686	2,728
Saudi Arabia	0.0	1.5	58	1,144	72
Singapore	6.7	7.8	1,164	51,121	0
UAE	-2.1	2.9	127	24,218	0
Ireland	3.2	7.3	1,118	119,569	1,226
Turkey	5.4	3.7	2,116	43,508	325
Egypt	5.4	4.6	1,108	–	536
Indonesia	6.1	4.2	123	42,503	0
Pakistan	6.3	3.7	232	–	–
South Korea	8.9	5.7	4,619	76,913	56,214
Argentina	-0.7	4.3	2,119	5,558	899

Source: The World Bank, *2002 Development Indicators*, Table 5.11.

Do Life Sciences Have a Future in the Arabic-Speaking World?

This is a region whose people and governments have in the past demonstrated extraordinary leadership ability in generating and understanding the knowledge sets necessary to succeed in the life sciences. Faced with another opportunity to succeed in this field, it would be a pity if it failed to pose the question "Should we become players in life sciences?" and simply watched this revolution in agriculture, chemicals, computing, mathematics, biology, energy and medicine pass by.

There are several essential components to the life sciences. Today, the first and foremost is biology. However, one could hardly do research and build a strong life science cluster without also having doctors, chemists, physicists, engineers, computer programmers, agricultural experts, lawyers and ethicists.

The recent explosion in life sciences knowledge was originally triggered partly by the interest of some of the world's great physicists. Niels Bohr, discoverer of key principles in atomic structures and radiation, wrote a seminal article in 1933, which laid out some key questions for life science research.[39] Erwin Schrödinger, another prominent physicist, followed this with a series of key questions in a short book entitled *What is Life?*[40] The book was so powerful that it changed the interests and careers of a series of prominent young scientists. Jim Watson, co-discoverer of DNA, began his studies thinking he would become an ornithologist. However, after reading Schrödinger, he began to study what would become molecular biology.

In short, success in life sciences depends on being able to attract, interest and apply some of the country's and the world's best minds across a series of disciplines. There are extraordinary Arabic-speaking science scholars and practitioners. However, more than a few, like Dr Farouk El-Baz, have made a global contribution from scholarly chairs in the United States and Europe.[41] Developing, attracting and putting this caliber of talent to work regionally is a crucial component of any country's future.

If a comprehensive peace were achieved tomorrow in the Middle East, would everyone's problems be over and everyone's prosperity assured? Or are current conflicts simply masking deep problems in terms of differentials in the ability to generate wealth?[42] Regional peace, if and when it comes about, may bring an end to armed conflict, but it does not end economic competition. Despite some individual success stories, parts of the Islamic world are falling way behind Europe, as can be seen from Table 3.5.[43]

Table 3.5
GDP Per Capita for Selected Countries (1970–99)

Countries	*1970*	*1980*	*1990*	*1999*
37 Islamic	US$197	US$1,145	US$1,071	US$1,124
19 OECD	US$3,123	US$10,673	US$21,516	US$28,028
%	6%	11%	5%	4%

Source: Bruce Scott, Jamie Matthews, HBS with data from World Bank, IMF, USDepCom.

The life sciences are no guarantee nor are they a panacea. There is no simple answer to the question, "Should they become a keystone of development for Arabic-speaking states?" Leading-edge technology development is expensive and risky (see Figure 3.7). Many companies, countries and regions have gone broke by betting on the wrong technology, or by betting too early or too late. However, the question of should we be generating more knowledge, in life sciences or in other areas of inquiry, should at least be asked and addressed.

Figure 3.7
Patents Granted by WIPO 1998

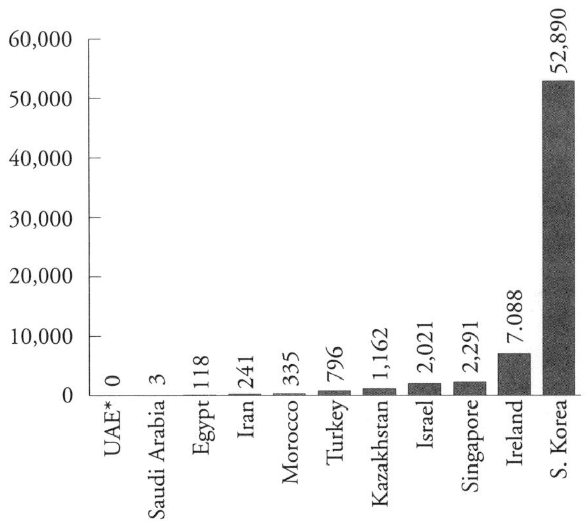

Country	Patents
UAE*	0
Saudi Arabia	3
Egypt	118
Iran	241
Morocco	335
Turkey	796
Kazakhstan	1,162
Israel	2,021
Singapore	2,291
Ireland	7.088
S. Korea	52,890

Y-axis: 0, 10,000, 20,000, 30,000, 40,000, 50,000, 60,000

Note: *No patents granted.
Source: With data from WIPO.

As the debate continues over where the region should go, let us return to the initial proposal made in this chapter. Perhaps the region is not conservative enough; it has forgotten some of the roots of its success. Sayyed Hossein Nasr quotes three important Islamic traditions in his book, *Islamic Science: An Illustrated Study*:[44]

The quest for Knowledge is obligatory for every Muslim.
Verily the men of knowledge are the inheritors of the prophets.
Seek knowledge from cradle to grave.

Section 3

ISSUES OF CONCERN

4

Stem Cell and Cloning Research: Implications for the Future of Humankind

John Gearhart

> *They would overcome flab with diet and exercise, wrinkles with collagen and Botox, sagging skin with surgery, impotence with Viagra, mood swings with anti-depressants, myopia with laser surgery, decay with human growth hormone, disease with stem cell research and bioengineering. They would banish germs with anti-fungal toothbrushes, hand-sanitizing lotions, organic food, bottled water, and echinacea.*[1]

An age-old dream of humankind has been to replace damaged, diseased or worn-out parts of the body with new, fully functional cells, tissues or organs. Recent advances in biomedical research may soon be able to transform this fantasy into reality. The discovery of stem cells from a variety of human tissues, both adult and embryonic, has opened a vast new terrain for scientific exploration in the field of basic biology. Embryonic stem cells, a controversial source of cells from which all cells found in the body can be formed, have been proved in animal experiments to produce functioning cells. This has led to the great expectation that these stem cells will eventually enable us to develop cell-based therapies for a variety of human diseases and injuries. In fact, we are now witnessing

.he dawn of a new field of medicine, regenerative medicine, which encompasses the development of therapies based on stem cell technology. Stem cell research is arguably the most important topic in biology today, focusing society's attention sharply on ethical, political and economic aspects of this cutting-edge technology.

Human embryo cloning, with applications for both reproductive cloning and 'therapeutic' cloning, is now intertwined with stem cell research. Hundreds of animals (mice, cows, sheep, pigs and cats) have been born that have been derived from the DNA of cells taken from an adult of the same species; that is, as a result of reproductive cloning. This procedure has led to a high percentage of abnormalities and unexplained deaths in the cloned animals and should not be performed with humans, although there have been claims of several human pregnancies with cloned fetuses. Therapeutic cloning – namely, the production of an embryo to be used for stem cell derivation – is another controversial procedure. It is now approved in the UK and is under heated debate in other countries. This would enable investigators to learn the mechanisms involved in a cell becoming a heart muscle cell in contrast to a nerve cell, for example. Such information is absolutely essential for developing cell-based therapies directly from patients' own cells. This is the medicine of the future.

Thus, stem cells represent one of the most promising areas of scientific research. These cells, which are not only involved in the formation of the tissues and organs during development, but also in the repair and replacement of cells in the adult, have become the focus of regenerative medicine and public policy in science. Although the field of stem cell biology has had a rather short developmental history, the concept of cells with the properties of what we now call stem cells has been the subject of investigations for almost a century. The term "stem cell," as referring to a cell capable of self-renewing and of specialization, was introduced in the 1970s in studies of hematopoiesis. Also during this period, other investigators were recognizing that a single cell within tumors known as teratocarcinomas had the capacity to form many different cell types found within the tumor as well as more cells like itself. The embryonal carcinoma (EC) cells were referred to as stem cells, both of the tumor and of the differentiated cells within the tumor. Over time, it was recognized that in adults, the further growth of tissues, or the repair of tissue injury, was undertaken by stem cells residing in tissues.

In the 1980s the derivation and culture of embryonic stem (ES) cells and embryonic germ (EG) cells from the mouse contributed immeasurably to our understanding of development in the mouse. The field of stem cell research was given a major boost with the successful derivation of ES and EG cells from human embryos in the late 1990s with the attendant promises for therapies developed from these cells. Since the embryo is destroyed to obtain these stem cells, ethical issues have divided the public over the production and use of these cells. Most recently, the evidence (controversial as it may be) that adult stem cells have a much higher degree of developmental plasticity or potential has heightened the excitement in stem cell research as well as influenced the debate on the use of embryonic and fetal sources of stem cells.

Pluripotent Stem Cells from Embryonic and Fetal Tissues

Of particular interest are stem cells that are pluripotent; that is, capable of forming all cell types present in the body or the developing embryo and fetus. Several tissue sources of these pluripotent stem cells are known in mammals.

As a distinct cell type, the pluripotent stem cell was first recognized in teratocarcinomas. These are bizarre gonadal tumours ("terat" means "monster" in Greek) containing a wide array of tissues derived from the three germ layers of the early embryo – the endoderm, mesoderm and ectoderm. These tumors may contain a large assortment of cell types including cardiac, striated and smooth muscle, cartilage, primitive neuroectoderm, various neuronal and glial cells, bone, glandular epithelia and occasionally, structures that represent small histologic sections of kidney, endocrine glands and so forth. Also present are very characteristic cells that are not differentiated, the embryonal carcinoma (EC) cells or the mixed germ cell carcinoma cells. The differentiated cells of the tumor are formed from EC cells present in the tumor. These EC cells are derived from primordial germ cells (PGCs), the embryonic precursors of the gametes. The EC cells can be differentiated, both in vitro and in vivo, into various cell types as well as giving rise to more EC cells. Tumors in which the EC cells have become completely differentiated are referred to as teratomas. EC cells are also one of the main components of human testicular germ cell tumors. Cultured EC cell lines were derived by isolating EC cells from tumors and growing them in a medium containing serum,

Figure 4.1
Sources and Derivation of Pluripotent Stem Cells

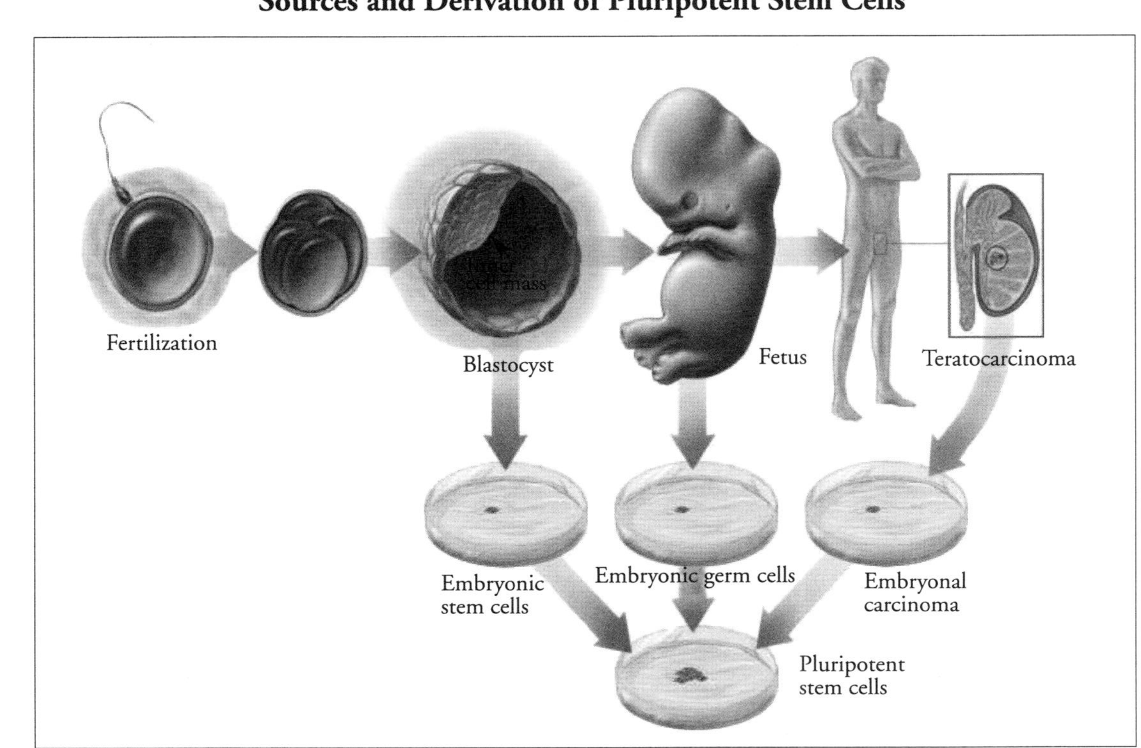

Note: Pluripotent stem cells have been derived from several human tissues: pre-implantation stage embryos, fetal, and testicular tumors.

Source: Michael S. Linkinhoker.

either in the presence or absence of a mitotically inactivated layer of fibroblasts, termed a feeder layer. In the 1970s and 1980s, EC cell culture and differentiation were popular experimental systems in which to study the differentiation of mouse and human cell lines. However, virtually all of these lines were chromosomally abnormal, raising concerns about the normality of the cell derived. The culture conditions developed for the EC cells proved critical to the derivation of embryonic stem cells that followed in the early 1980s.

In 1981, embryonic stem (ES) cells were derived from pre-implantation-stage embryos of mice. Evans and Kaufman and Martin derived pluripotent stem cells from the outgrowth of cultures of mouse blastocysts on feeder layers. The outgrowths consisted of different types of cells, some of which remained undifferentiated. The undifferentiated cells could be subcultured and expanded to form established cell lines that could give rise to more of the undifferentiated cells and to a variety of differentiated cells. Martin termed these cells embryonic stem cells. The fact that these appeared to divide indefinitely led investigators to refer to these cell lines as being immortal.

Mouse ES cell lines have had a major impact on many areas of science during the past two decades: developmental biology, genetics, tumor biology and behavior – the list is extensive. Much of this impact was due to the use of ES cells in genetic manipulations with the subsequent production of whole animals for study. In vitro differentiation of ES cells developed into a small industry, with emphasis initially on neural, cardiomyocyte and hematopoietic derivatives.

With the importance of mouse ES cells established, investigators tried for years to obtain human ES cells – a goal first achieved by Jamie Thomson in 1998.

Human ES lines are derived from the inner cell mass (ICM) of blastocysts obtained through in vitro fertilization and donated by individuals through informed consent for research. Subsequent to Thomson's success, several laboratories have now established human ES lines.

Embryonic germ (EG) cells are derived from cultured PGCs, the same cells from which EC cells are derived spontaneously or experimentally in vivo. In the presence of serum and certain growth factors, mouse PGCs isolated directly from the embryonic gonad onto feeder layers will form colonies of cells that seem morphologically indistinguishable from EC cells or ES cells grown on feeder layers.[2] In 1998, our laboratory first reported the derived EG lines from PGCs of human cadaveric tissue.[3]

Figure 4.2
Embryonic Stem Cells

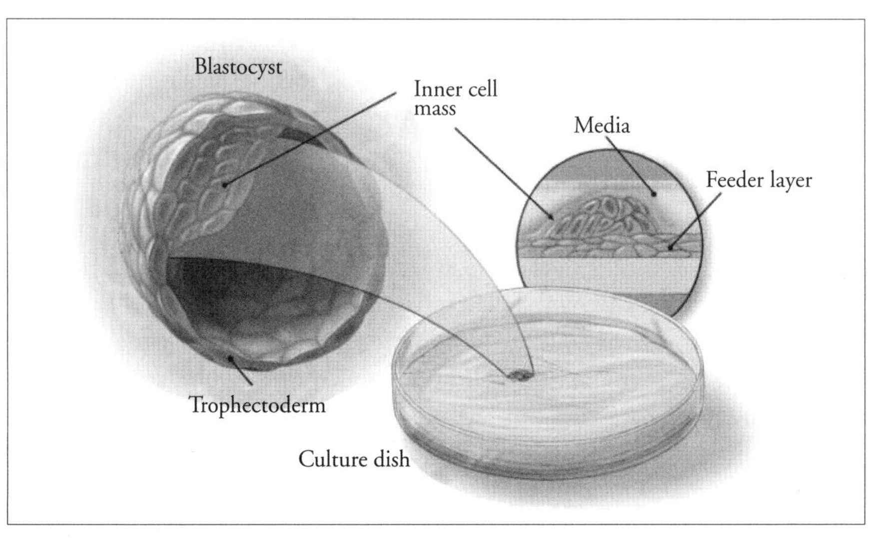

Note: Cells from the inner cell mass of the blastocyst are isolated and cultured, giving rise to embryonic stem (ES) cells.

Source: Michael S. Linkinhoker.

Figure 4.3
Embryonic Germ Cells

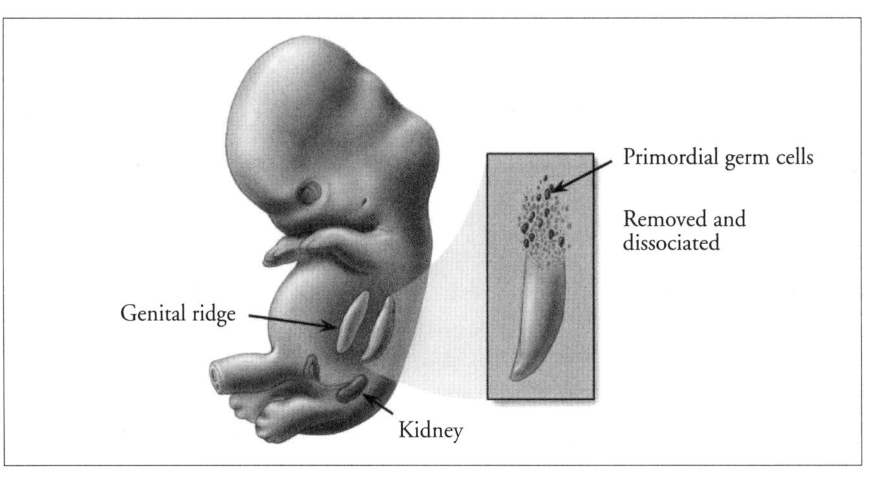

Note: Primordial germ cells are isolated from the developing gonad (ovary or testis) and placed in culture, giving rise to embryonic germ (EG) cells.

Source: Michael S. Linkinhoker.

Although EC, ES and EG cell lines have been isolated from mice and humans, only ES cells have been isolated from non-human primates. Stem cell lines vary, even within species, in some properties and information obtained, and therefore protocols and procedures developed for one species may not be appropriate for stem cells from other species. Growth-culture requirements, cell-cycle times and cell-dissociation conditions have all been found to differ. New procedures for efficiently transfecting (genetically manipulating) the human ES/EG lines had to be developed. All this indicates that research must be performed directly on any given source of stem cells and that some technologies cannot be directly transferred among sources.

The pluripotent stem cell lines have many common attributes, with some exceptions of uncertain significance. Some of the classical markers of these cells include an isozyme of alkaline phosphatase, the POU-domain transcription factor Oct4,[4] high telomerase activity and a variety of cell-surface markers recognized by monoclonal antibodies to stage-specific embryonic antigens or to tumor-recognition antigens. Although some of these markers are not unique to stem cells, they can nevertheless serve as reagents with which to physically separate pluripotent stem cells from their differentiated derivatives. The physiological significance of most of the markers is unclear, with the exception of Oct4. Compelling studies carried out in mouse EC, ES and EG cells, as well as in mouse embryos, point to a critical role for Oct4 in the establishment and/or maintenance of pluripotent cells in a pluripotent state. Differentiation of pluripotent cells is associated with downregulation of Oct4 levels, and downregulation of the Oct4 gene in ES cells or in mice results in the differentiation and loss of pluripotent cells.

Table 4.1
Features of Pluripotent Stem Cells

	ES	*EG*	*EC*
Self-renewal	positive (+)	+	+
Karyotype	normal	normal	hyperdiploid
Oct 3-4	+	+	+
Telomerase	high	high	high
Hoescht exclusion	+	+	+
Imprint	+	erased	not determined
Tumor form	+	+	+

Source: "Nature Insight: Stem Cells", *Nature*, 2001; Marshak et al., 2000.

Developmental Potential

The developmental potency of mouse pluripotent stem cells has been tested in three independent assays:

- in vitro differentiation in a petri dish;
- differentiation into teratomas or teratocarcinomas when placed in adult histocompatible or immunosuppressed mice;
- in vivo differentiation when introduced into the blastocyst cavity of a pre-implantation embryo.

All of the pluripotent stem cells can differentiate in vitro into a wide variety of cell types representative of the three primary germ layers in the embryo. When pluripotent stem cells are introduced into histocompatible or immunocompromised mice, they form tumors that are indistinguishable from the gonadal tumors from which EC cells were originally derived. In chimaeras, mouse ES and EG cells contribute to every cell type, including the germ line. In contrast, murine EC cells introduced into embryos colonize most embryonic lineages, but generally do not colonize the germ line, with one experimental exception. The inability of EC cells to form functional gametes most likely reflects their abnormal karyotype. Because of ethical concerns, non-human primate and human pluripotent stem cells have not been tested for their ability to participate in embryonic development in vivo, but in the other assays, such as differentiation in culture and tumor formation in immunocompromised mice, they behave identically to their mouse counterparts.

In Vitro Differentiation

> We are standing and walking with parts of our body which we could have used for thinking if they had been developed in another position in the embryo.[5]

The ability of pluripotent stem cells to give rise to a wide array of differentiated derivatives is, of course, the reason why they may be so useful for purposes of cell-based therapy.

Figure 4.4
Challenge for Directing Stem Cell Differentiation

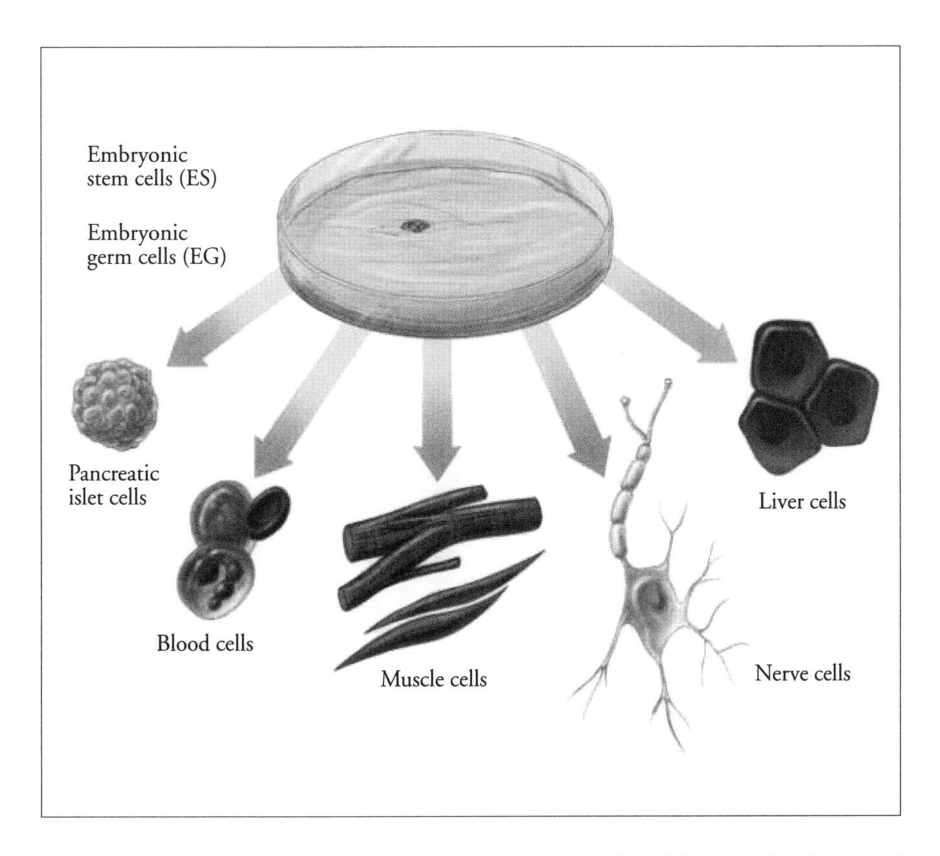

Note: The pluripotent stem cells (EG/ES) in culture are capable of forming all cell types of the body. The experimental challenge is to direct the stem cells to produce only one desired cell type.
Source: Michael S. Linkinhoker.

In the absence of factors that inhibit their differentiation, pluripotent stem cell differentiation has typically been directed by manipulating their environment by trial and error, in attempts to mimic the environment found in normal development within the embryo. This can be achieved by growing stem cells at high density, by growing them on different types of feeder cells, by addition of growth factors, or by growth on crude or defined extracellular matrix substrates.

Figure 4.5
Differentiation of Stem Cells Through Different Culture Conditions

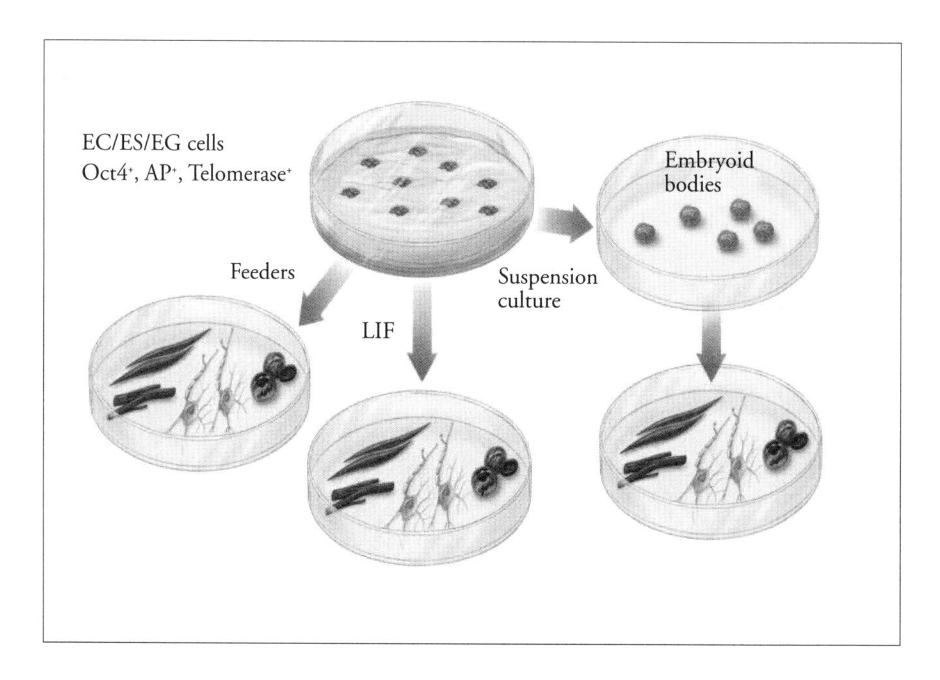

Note: Differentiation of the stem cells can be initiated through changing the culture conditions. These changes include the growth of the cells in suspension (as opposed to their growth on feeder layers that prevent cell differentiation), the removal of feeder layers, or changes in the growth factors in the media.
Source: Michael S. Linkinhoker.

In these conditions the types of differentiated cell types formed can be varied and haphazard. When grown in suspension, pluripotent stem cells will form embryoid bodies, structures originally described in teratomas and which resemble the early peri-implantation embryo. In forming embryoid bodies, stem cell differentiation may proceed in a way related to that occurring in the embryo. Many differentiated cell types can be derived, including neurons, glia, cardiomyocytes, skeletal myocytes, adipocytes and hematopoietic cells. Differentiated cell types must be sorted out from each other and away from stem cells. This can be carried out by traditional methods of fluorescence-activated cell sorting (FACS) if suitable cell-surface markers are available, by selective growth of differentiated cells if suitable culture conditions are known, or by introduction of a selectable marker that allows either FACS separation of differentiated cells or drug selection to ablate stem cells and other unwanted cells.

Figure 4.6
Selection Schemes to Isolate Differentiated Cells

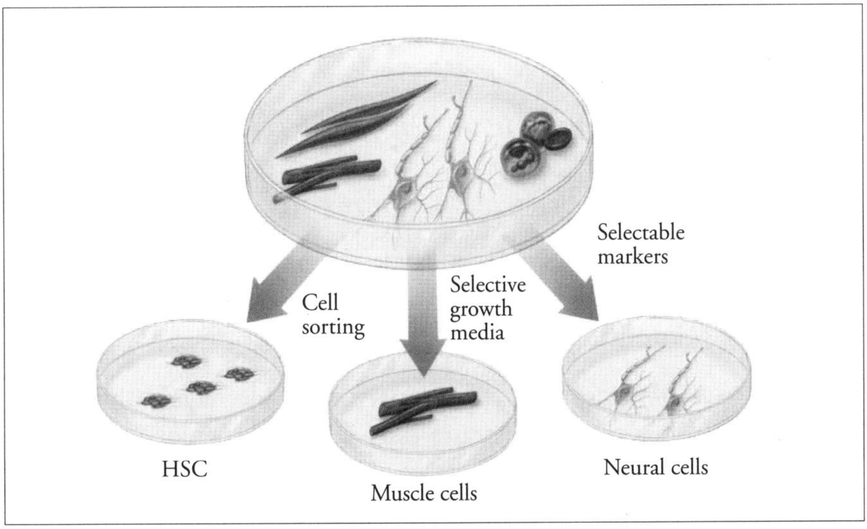

Selectable markers
Cell sorting
Selective growth media
HSC
Muscle cells
Neural cells

Note: Once differentiation of the cells has been initiated, selection schemes are used to isolate specific differentiated cells. These include sorting cells for proteins on their surfaces, growing them in conditions that will support a specific type of cell, or using genetic manipulations of cells that can be used to selectively permit the growth of specific cell types.

Source: Michael S. Linkinhoker.

Do the results of transplantation studies in animal models suggest that the ES/EG stem cells can be used as a source of cells to ameliorate or repair diseased or injured tissues? There are an increasing number of reports with mouse ES cell-derived cells transplanted into mouse or rat models of diseases or injuries. These include some of the most debilitating diseases such as spinal cord injury,[6] Parkinson's disease,[7] immune deficiency[8] and diabetes.[9] The results of these studies, referred to as proof-of-principles studies, are all consistent with the idea that cell-based therapies can be effectively developed for many disabilities. This does not mean that we fully understand the basis on which the animals have recovered, nor the most effective means of using the cells for therapies, nor that this approach is without risk. However, it is encouraging that the animals show improvement following the grafting of cells and that there is now proof that the cells will function within the animals. Few such experiments have been performed with human cells but this reflects the fact that the human cells have only been available a short time. There are unpublished reports of human ES/EG cell-derived proof-of-principles experiments paralleling those mentioned above.

Current Human Cell Research Focus

With research progress being made in many areas of stem cell biology, emphasis has been placed on certain aspects, as discussed below.

Cell Culture Conditions

The lines in current use have been derived on feeder layers of cells that provide growth factors or substrates essential for the establishment and maintenance of the stem cells. These feeder layers were generally derived from mouse embryonic tissues and used for mouse stem cells culture before their use with human cells. There is now emphasis on developing human feeder layers, 'better' feeder layers, or feeder-layer-free cultures of human cells.

Interspecific feeder layers are of concern as they may transmit endogenous viruses to the human cells, placing recipients at risk. Indeed, the FDA considers any human cells grown in the presence of other animal cells or products as xenografts and subject to more rigorous requirements when intended for use in humans. It is therefore desirable to have human cells as a feeder layer to avoid the safety concerns regarding cross species transmission of endogenous viruses. Since the culture of human stem cells on non-human feeder layers requires that such xenografts when used for human transplantation fall under more rigorous FDA tests, it would be best to avoid this possibility altogether.

The search for "better" feeder layers has always been a part of the ES/EG field. "Better" is generally defined as allowing for robust cultures with no or little evidence of spontaneous cell differentiation. These goals are relative and it is not clear what the endpoints should be. For example, selection occurs during culture and faster-dividing cells may lose some of the desirable properties of the stem cells. It is not unusual, inadvertently, to generate a number of sub-lines of cells during culture with altered abilities to differentiate, grow in suspension, and so forth.

The presence of feeder layers in cultures confounds studies of the stem cells in culture in that there is always the background of the cells to deal with. Efforts are being made to determine what the bases of the feeder-layer requirements are; that is, are they necessary for physical attachment, do they produce specific growth factors, or both, that allow for the growth of the stem cells? It should be possible eventually to coat plates with the desired physical substrate and provide the factors either in the media or attached to the substrates to replace the feeder cells. Robust feeder-layer-independent cultures would represent a tremendous advance in this field.

Most culture conditions employ the use of animal sera in the media. This becomes problematic experimentally in efforts to control or regulate the differentiation of the cells, as sera vary, even though selected for ideal ES growth, from batch to batch. It is important to develop completely defined media (that is, without sera) with the necessary factors added exogenously so that we can discover the role of various factors and their combinations in cell differentiation. Determining the critical factors in sera is, therefore, a focus of research.

Derivation of New Lines

New lines are continuously being derived for a variety of purposes. For example, since all of the current lines being used were derived with mouse feeder layers, new lines must be produced on human feeder layers for them to be free of concerns regarding viral transfers. Some investigators feel that the current material transfer agreements (formal agreements setting the conditions for obtaining cell lines from other investigators or institutions) are too strict and desire to have their own lines. Some investigators feel that the current lines do not represent sufficient diversity and thus that new lines are needed. Others are concerned about the continued quality of the existing lines; that is, accumulation of mutations (see below), and feel that new lines are desirable. In the United States, any lines produced after 9pm, August 9, 2001 are ineligible to receive federal funding.

Property of Stemness

As mentioned earlier, stem cells have the properties of cell division and cell differentiation, referred to collectively by some as "stemness." Over time, a listing has been compiled of the characteristics of ES/EG cells, based on several antigens, biochemical markers, some expressed genes, and measures of developmental potential. There is now a push to define "stemness" in molecular terms, through genomics and proteomics. Thus, many laboratories are now determining gene expression profiles and protein profiles for their various stem cell lines and comparing among them to find the sets of genes that could potentially define stem cells. This information is important for rapidly determining whether any new lines are actual stem cells, for monitoring the specialization or differentiation of cells into defined lineages, for monitoring the reprogramming of differentiated cells or nuclei to a more embryonic state, and eventually for engineering cells to become stem cells.

Development of High-Efficiency Differentiation Protocols

Major advances in cell biology, developmental biology and molecular embryology have revealed to some degree the mechanisms underlying cell differentiation. We now utilize this information in attempts to influence, enhance, direct or select for the differentiation of stem cells in culture. This is routinely done by controlling the growth environments, as mentioned earlier, or attempting to control the expression of genes that are critical in initiating a specific cell differentiation lineage or in controlling/regulating stages during the differentiating events. To date, although it is possible to significantly influence the differentiation of specific cell types, particularly within the central nervous system, for example, we are a long way from controlling these events to assure the production of a single cell type in large numbers. In other lineages, we are struggling to achieve any success in enhancing the differentiation of cells, as with hepatocytes, for example.

In developing these protocols for generating desired cell types in culture, an intriguing question is whether we must rely on recapitulating the events of normal development or whether there is more than one way to obtain cells with the desired functions.

Authenticity of Derived Cells

In the strategies used to obtain differentiated cells in culture from the stem cells, one major concern is whether the cells produced are authentic. In other words, do they possess the same properties as the cells going through normal development? What parameters for normal function should be assessed in the culture-derived cells? Is it important that the cells derived be identical to their normally derived counterparts? Do assessments in the dish provide any measure of how the cell will function when transferred into an organ or tissue? We are slowly learning some of these answers through empirical methods, but this process will obviously take time.

Various Stem Cell Sources: Comparison and Contrast

Of major interest to stem cell biologists (and the policy makers) is the comparison among stem cells sources in a number of aspects. Above all, the question of development potentiality (that is, how many different cell types a claimed stem cell source can produce) dominates the current scene. Little attention is given to whether the cells are functional, whether the variety of cell types derived came from a single cell (as opposed to a starting population with a mixture of cells), the frequency with which they can be generated, and whether or not the cells can be grown or expanded in cultures. In fact, all of these questions are relative to the utilization of these sources for therapies and will be reviewed at the end of this chapter. It is important to test various stem cells side by side to determine which are going to prove most efficacious for our desired goals. This aspect is now gaining more attention in laboratories.

Stem Cells and Stem Cell Lines: Mutation Rates, Frequencies and Types

As cells divide, gene alterations (mutations) occur at a set frequency and these mutations accumulate. If cells are cultured for hundreds of population doublings, the 'genetic load' in these cells will eventually impact on the stem cell properties of the cells. This could lead to lines not differentiating appropriately or to changes that could make them tumorigenic upon transfer. In one report on mutation frequencies and mutation types occurring in mouse ES cells, it was found that the mutation frequencies were lower than somatic cells but that the most common type of mutation found resulted in the loss of tumor suppressor genes.[10] Studies of this kind must be performed on human lines and, if similar to the mouse results, could underscore the need for new line derivations.

Grafting Studies for Testing Cell Function and Assessing Safety Concerns

Animal grafting studies are necessary to test the functioning of derived cells in situ and to assess the grafted cells and grafting strategies for safety issues. While animal models of human diseases and injuries may not precisely model the pathogenesis of the human condition, they are certainly the best testing system for determining whether cells derived for cell-based therapies function and whether they appear to be safe for human use. A clear distinction must be made regarding the purpose of such tests. Are the grafting experiments designed to test the ability of cells to differentiate in a developmental model? For example, this could mean placing the human cells in an animal undergoing development (that is, in tissues of a fetus or neonate), depending on the tissue. Alternatively, are the cells being placed in sites of injury or pathology in order to restore a function? The demands on the cells are quite different in each of these paradigms or models.

For each model, the stage of cell differentiation ideal for grafting must be determined. Should precursors, progenitors, or partially or fully differentiated cells be grafted? What is the best modality for delivery of the cells or targeting them to the appropriate sites? The answers to these questions will be determined empirically and may differ significantly for each disease/injury model system.

Safety Issues

The greatest concern in applying stem cell technology to patients is safety. Cells introduced into patients must not only provide a therapy but must also not compromise the patient by performing inappropriately. Such safety concerns arise as a result of some of the very features that make stem cells so attractive for cell-based therapies. The ability to divide indefinitely in a dish is a valuable feature for growing a large number of cells for experiments to study cell differentiation, genetically modify cells or produce sufficient cells for grafting experiments. However, this could have dire consequences following transplantation to patients. As mentioned earlier, ES and EG cells, when transplanted 'ectopically' in mouse studies, will form rapidly growing tumors – the teratocarcinomas. These tumors have been reported in early studies of grafts in mice, following flawed strategies to select only those neural cells that were derived from the stem cell. Included in those transplanted cells were the actual stem cells. Safety studies must be performed to demonstrate that the strategies used for producing the valuable populations of cells for cell-based therapies are devoid of any tumor-producing cells. Another possible strategy to ensure further against the introduction of such cells would be a fail-safe procedure by which cells could be genetically manipulated to die if they continue to express (or re-express) specific genes that are only expressed in the stem cell.

Other safety concerns include the appropriate and complete differentiation of the grafted cells and the localization of the grafted cells to the target site. It is difficult to predict how cells will respond when placed in an environment that is grossly different from the cell culture plate or a 'normal' environment in a tissue. The purpose of these grafts is to restore function within a diseased or injured site – an environment about which little is known. As mentioned earlier, environmental niches provide growth factors and signals that are critical not only for permitting or instructing cells with regard to specialization, but also for their survival. Sufficient tests of the cells in grafts must then be performed to ensure that the cells behave appropriately.

As mentioned earlier, the modalities of delivery of the grafted cells to the target site must be investigated to ensure that viable cells are introduced without undue trauma. However, once cells are introduced, they may indeed migrate from the target site, setting up a potentially dangerous situation either by differentiating in a new direction as a result of finding themselves in a new environmental niche, or continuing to differentiate in the desired lineage but in a new location in which such cells would be detrimental.

For all the above-mentioned reasons, safety studies are absolutely necessary. Regulating bodies in this area of medicine have yet to decide the extent of experiments to be performed to assure safety. These decisions will most likely be disease- or injury-dependent, as the pathological consequences and risks to the patient must be evaluated against the benefit. In the experimental phase of this work, investigators must be prepared to track all cells introduced into animal models for extended periods of time. This has now turned research emphasis to the development of 'labels' for the grafted cells that must persist for months (or years); that must allow for single cell resolution; that will not be toxic to cells; and that may allow for the real-time imaging of the cells, obviating the necessity to work with very large numbers of animals in order to have sufficient time points for analyses at necropsy. The safety studies will also fuel debates on the appropriateness of various animal models, the length of time that grafts should be evaluated and whether non-human primate studies should be required.

It is anticipated that the studies necessary to demonstrate safety alone for cell-based therapies will be expensive and will require time. Thus, even if appropriate cells are available, the clinical application of such therapies is still years away and would entail great expense.

Avoiding the Immune Response to Grafted Cells

After grafting, cells and tissues must not only integrate and function, but must also avoid rejection by the host's immune system. Unless the cells originally came from the patient or a close relative, all grafts will elicit an immune response. Several strategies can be employed to avoid immune detection of grafted cells obtained from ES sources. These strategies are discussed later and include suppressing or altering the host immune response through the use of drugs or pre-treating the patients to induce tolerance to the grafted cells; genetic alterations to the cells to make them less likely to evoke an immune response; or the use of a new procedure called somatic cell nuclear transfer in which embryonic stem cells with the patient's genes are derived and used to obtain the cells for grafting.

Somatic Cell Nuclear Transfer (SCNT) and Stem Cells

The technique of somatic cell nuclear transfer (SCNT) will be discussed at length since it is currently under keen debate and has great practical impact on research. SCNT has also been referred to as therapeutic cloning (as opposed to reproductive cloning), embryo cloning, research cloning and nuclear transfer research.

Figure 4.7
Somatic Cell Nuclear Transfer (SCNT)

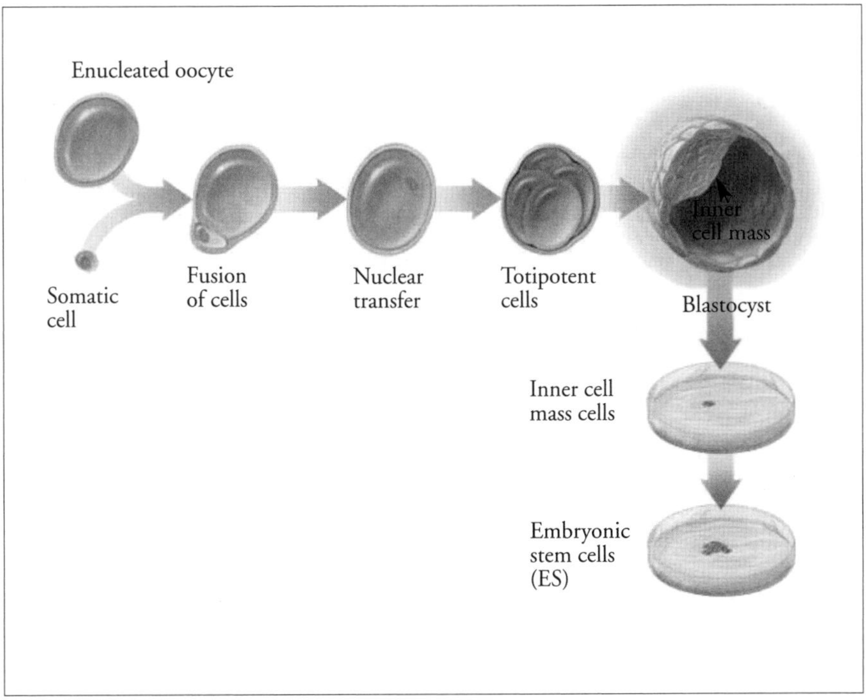

Note: In somatic cell nuclear transfer (SCNT), the nucleus of a cell from a patient is introduced into an egg from which the original nucleus has been removed. The reconstituted cell undergoes the early developmental stages of embryogenesis, giving rise to a blastocyst from which the inner cell mass is isolated and stem cells derived, as in Figure 4.2. The resulting stem cells have the same nuclear genome as the patient.

Source: Michael S. Linkinhoker.

Any tissue derived from embryonic stem cells will express histocompatibility antigens and, therefore, will be rejected if grafted into a genetically disparate recipient. This is a major obstacle that needs to be overcome if the therapeutic promise of stem cells is to be realized. Theoretically, there are several options that can be considered to overcome the histocompatibility problem: lifelong immunosuppression, the induction of transplantation tolerance, large stem cell tissue banks for matching, genetic alterations (targeting) of the histocompatibility genes in the ES/EG cells, or somatic cell nuclear transfer. Immunosuppressive drugs are problematic, causing serious complications with life-threatening possibilities. Many strategies for inducing graft tolerance have not yet proved efficacious. The possibility of deriving selected stem cell lines with the appropriate genotypes/haplotypes for matching large segments of a population, although theoretically possible, would require hundreds of cell lines to cover just 75 percent of the population. The expense and logistics of this endeavor would be daunting. Indeed, this is a major argument that is made by the advocates of umbilical cord blood sources of multipotential stem cells for the use of this source, as cord blood storage is growing into a major industry with hundreds of thousands already stored. The key issue here, of course, is how valid are the claims for multipotentiality of cells obtained from this source. Conducting genetic alterations of the major histocompatibility genes, either to introduce a patient's DNA sequences into the stem cells or to alter them so as to make the cells less immunogenic, is currently a dream.

In summary, the histocompatibility issue will severely limit the use of ES/EG-derived tissues to a very, very small group of patients – those who would genetically match the existing lines available for therapies, and those who could tolerate lifelong immune suppression. Those who advocate using stem cells derived from adult tissues stress this point, since it is the major limitation of embryonic sources of stem cells.

It is no wonder, then, that the use of ES lines derived from somatic cell nuclear transfer, in which the cells have the genetic identity (nuclear genes) of the donor (or patient) has received such attention. There are now many other terms used to delineate this SCNT procedure, some with negative connotations – therapeutic cloning, research cloning, regenerative cloning, nuclear transfer and nuclear transplantation.

Indeed, proof-of-concept for this approach has been reported in mouse and bovine studies. Tissues generated from ES cells following SCNT were not rejected when introduced as grafts. Imagine the following scenario. A patient, diagnosed with any illness (from a range of non-immediate life-threatening illnesses), could have a noninvasive tissue biopsy (buccal smear) performed to obtain cells, have the nuclei individually introduced into enucleated eggs, and have ES lines derived from the resultant blastocysts. Whatever the cell type needed by the patient, it would then be grown in culture from the stem cells, and introduced into the patient for therapy.

While the immune rejection issue has been used as the strongest argument to promote the use of SCNT, there are several other arguments that are important for the SCNT technique.

In a culture sensitive to diversity issues, it is important that cell lines from genetically diverse populations be included in the stem cell studies if all segments of the population are to benefit from stem cell research. The use of assisted reproductive technologies (ART) is widespread and is used as the source of donated embryos for stem cell derivations. However, ART utilization is representative of a very narrow population with respect to genetic, racial and ethnic diversity. SCNT could be used effectively to generate cell lines with wide diversity that could prove useful for tissue matching and for uncovering differences in the variability of genetic mechanisms in disease and developmental biology.

SCNT would be an effective means by which stem cell lines could be generated to facilitate the study of genetic diseases. Genetic factors have been identified that increase the risk for some diseases but only those having the disease possess all the genetic variants for the disease. SCNT could be used to develop valuable lines and reagents in studying several diseases that stem from a combination of factors. These include neurogenerative diseases, multiple sclerosis, diabetes and cardiovascular diseases. Such studies would not only involve the pathogenesis of the disease but also help to identify all the genetic factors involved in the disease process. An interesting use of this approach has been demonstrated in the mouse, where the gene defect for an immunodeficiency was corrected in stem cells following SCNT, and the corrected stem cells were used to rescue the defective mouse from which the donor cells were obtained.

Genetic causes of human developmental anomalies, particularly those affecting early development, are currently difficult to study since access to human fetal tissue with the mutations is rare. Through the use of SCNT, lines could be generated that would be valuable in analyzing the mechanisms of action during the developmental stages of tissues and cells.

Somatic cell mutations play a critical role in all human cancers as well as in some other diseases. Nuclei from the diseased cells of the patients could be the source of stem cell lines via SCNT. Through the use of these lines, we would be able to get valuable insights into how the mutated or altered genes function and how they cause the healthy cells to form cancers.

In the long term, I believe that the SCNT studies will lead to fundamental discoveries that enable reprogramming of any adult cell into stem cells, which then can be differentiated into desired cell types for therapies. As evidenced by the number of cloned mammals reported, whether completely normal or not, it is clear that the egg cytoplasm contains factors that are responsible for the genetic reprogramming of nuclei from differentiated cells into embryonic cells. Once we learn what these factors are, and perhaps how they work, we should then be able to reprogram cells, either in the dish or in vivo.

In summary, it may be noted that the potential medical benefits of SCNT/stem cell technology are enormous. As it is impossible to predict the direction of science in the short or long term (considering the general naivety in assessing the impact of fundamental discoveries), one cannot judge the real magnitude of SCNT benefits. In the debates concerning SCNT, little attention has been paid to the importance of generating intellectual property (IP) in this area, the development and use of new technologies around this IP, the effect of this research on mobilizing scientific teams to focus on projects, the inducement of the next generation of scientists into this field, and the investment of money and resources.

In view of the potential benefits of SCNT research with human cells, what are the real concerns? Before this scenario is extended too far, it is best to look at some practical concerns, both biological and societal. From the experimental perspective, it is not yet known how difficult it will be to produce blastocysts following SCNT with human eggs. The frequency of success will impact substantially on this area of research, as the availability of human eggs and the issue of consent are matters of concern. There have been reports in the press of successful human embryo production but these have not yet been corroborated in peer-reviewed journals, so this data cannot be evaluated.

With respect to the feasibility of avoiding the immune response to grafted cells, some practical issues remain. There are minor histocompatibility antigens (genes) in rodents that turn out to be mitchondrial in origin. Proteins coded for by genes in the mitochondria are expressed on the cell surface and will lead to graft rejection in rodents. SCNT only involves the nuclear genes so it is possible that mitochondrially coded antigens could lead to graft rejection of the human cells.

Also, the timeframe for deriving stem cells from SCNT embryos, and the production of differentiated cell types for grafting to the patient could take a year, thus limiting the types of diseases or injuries appropriate for cell-based therapies. This estimate is based on current protocols in this area, with the relatively long cell division time of human cells from 34 to 36 hours. This timeframe could also be lengthened by any requirements for safety studies or quality control, which could be imposed by the Food and Drug Administration, for example.

Opponents of SCNT raise several ethical issues that they feel should be sufficient to preclude this research. There is great concern that any technological advances made due to new knowledge gained through discovery will be used to enhance the production of SCNT blastocysts and increase the likelihood of the successful cloning of human beings. This is the slippery slope argument that obscures the fact that there is a very broad boundary here that must be respected; that is, no embryo produced with SCNT should be returned to the uterus. With SCNT, embryos would be produced specifically to be destroyed in contrast to the more acceptable use of extra embryos generated for childbirth through assisted reproductive technologies. A further argument, however, is that women will be exploited for their eggs and a market for human eggs will be developed, as the demand for eggs intended for research purposes will be in the millions. To opponents, these arguments take on added significance, since they feel there are alternatives: stem cells from adult tissues or from umbilical cords.

The Controversies in Stem Cell Research

There is overwhelming scientific opinion that stem cell technology will enable us to develop new forms of therapies for diseases and disabilities. Results in animal studies are consistent with this notion. However, there are controversies in stem cell research that must be acknowledged and addressed for reflecting societal responsibilities. It is not the purpose of this review to present pro and con arguments with respect to these controversies but only to introduce the topic since these concerns impact directly on the progress of research in this area.

Many countries are now struggling with these controversies, and the resolutions are as varied as the world cultures. The controversies include the sources of stem cells, although every scientific review of therapeutic opportunities afforded by adult-derived and embryonic stem cells has concluded that the embryonic stem cells are far more versatile for medical therapies.[11] Should there be public funding for embryonic and fetal sources? Should the cloning of human embryos for research purposes be permitted? Should there be ownership of the cells and technology derived from them? How should all of these areas be regulated so as to prevent abuses? For example, will research with SCNT lead to discoveries that will enable us to clone embryos more effectively, in which case this could lead to attempts to clone human beings. Although the boundary between these two purposes is very broad and unambiguous (that is, no SCNT-derived embryos should be transferred to a uterus), there is concern that it will be breached.

In Western cultures, with our multi-ethnic, pluralistic societies, any resolution of these controversies must reflect the differences in the moral interpretations of early stages of human development and the moral obligation to help those who suffer. We must develop ethical standards that are endorsed by the majority and develop sound research regulations that will permit the advancement of science while recognizing the legitimate alternative views – daunting but necessary tasks.

The Future

> The health of the people is really the foundation upon which all their happiness and all their powers as a state depend.[12]

As a result of recent advances in cell biology, developmental biology, molecular biology and the completion of the Human Genome Project, we are now in a position to understand how cells really work. The problems are formidable but the potential benefits are great, both for those who seek to understand biology and for those seeking to find therapies. Stem cells are telling us how the instructions are laid down for making a tissue and organs, and indeed, the whole organism! Also, how these instructions can be reversed, and how we might reconstitute them, for repair or replacement of damaged or diseased tissues and organs. Stem cells will enable us to discern many principles of biology and provide us with an unprecedented opportunity to study all aspects of human development.

For the near future, stem cells will be at the center of the developing technology for cell-based therapies. The stem cells will serve as the direct source of cells for transplantation either as specific cells, such as heart muscle, brain cells, and so forth, or through tissue engineering, organs or parts of organs (see Figure 4.8) and as the source of information to be applied to the patients' own cells, either through reprogramming of differentiated cells from adult tissues or the manipulation of their own stem cells in situ; that is, in the body (see Figure 4.9). We must, however, be circumspect, as stem cells will not be the answer to all of our medical ills. The stem cell promise, as presented in the Dowd quote at the very beginning of this chapter, will not extend to all ailments, and possible therapy will take years to achieve.

Figure 4.8
Cell Therapies: Transplantation

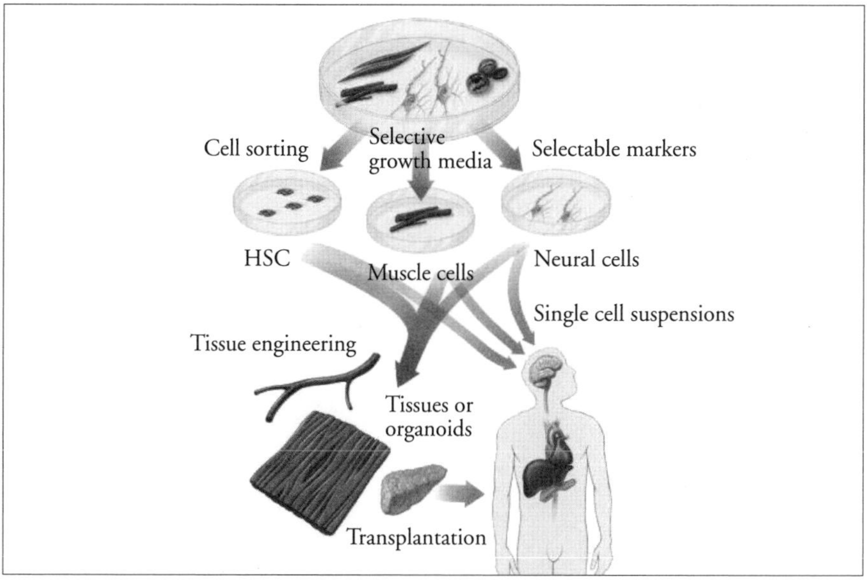

Note: Cell therapies will include the transplantation of specific types of cells such as insulin-producing cells, liver cells and nerve cells, as well as tissues and organs (or parts of organs) constructed from several types of cells.

Source: Michael S. Linkinhoker.

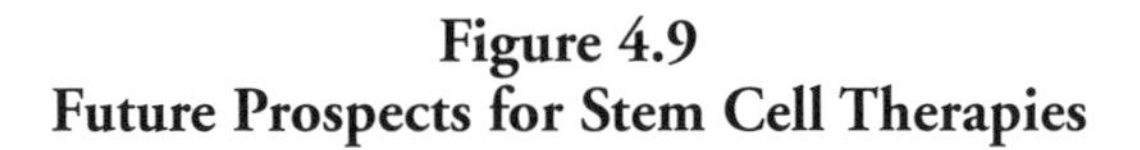

Figure 4.9
Future Prospects for Stem Cell Therapies

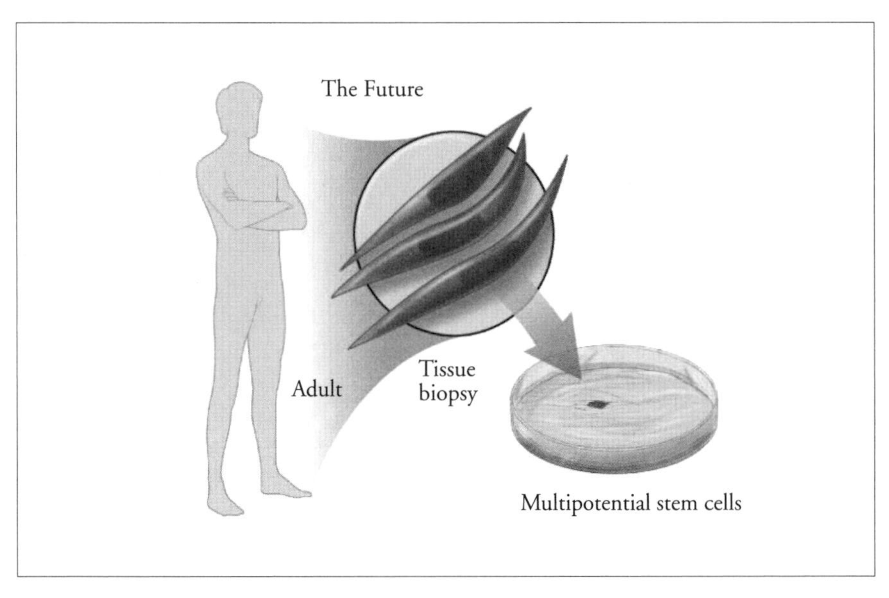

Note: It will be possible in the future to convert many types of cells from a patient into a stem cell and derive the type(s) of cell the patient needs for any given therapy.
Source: Michael S. Linkinhoker.

Current research in this field will be focused on fundamental discovery, since the scientific challenges in providing safe cell-based therapies are immense. It will take the combined efforts of many scientists and clinicians in a variety of disciplines to bring this endeavor to fruition. Clearly, with the advent of stem cell biology, the future ain't what it used to be!

5

Ethical, Legal and Social Issues in the Field of Biotechnology

Glenn McGee

It is, perhaps, the most important scientific advance in the past one hundred years, with vast unrealized potential. It stirs heated debate on issues ranging from abortion, cloning, fetal tissue, transplantation, gene therapy, animal rights and enhancement technology. It raises worries about women in research, sex, the regulation of in vitro fertilization (IVF) clinics, the danger of changing the human germ line, and the war against aging. Before it is developed, some of the most powerful politicians on earth will find themselves forced to modify deeply entrenched views and a few dozen scientists will become billionaires through patents on bits and parts of embryos. More than 150 million in the Middle East alone, and perhaps another billion around the world, may be treated with it before the decade comes to an end, yet almost no significant research involving human subjects has yet been performed with it. It commands the attention of the major newspapers, news media and scientific and business press every day, yet not a single book has been written about it. "It" refers, of course, to the human embryonic stem cell, perhaps the most important breakthrough in humanity's prolonged quest to understand its own origins, and key to dozens, perhaps hundreds of advances in medicine. It is also the most controversial technology imaginable, viewed by many as a Faustian bargain that involves the trading of innocent potential lives to extend other lives, and by many more as a necessary partial sacrifice of the human procreative mystery in the interest of curing disease.

Origins of the Stem Cell Therapeutics Debate

Dr John Gearhart and Dr James Thomson, along with perhaps one hundred other senior and junior researchers at the Johns Hopkins University, the University of Wisconsin and a dozen other universities, had raced to identify the *pluripotent* human embryonic stem cell for several years prior to the publication of their findings in 1998.[1] Long before any clinical demonstration that human embryonic stem cells (hES) could have therapeutic efficacy in the treatment of human disease, many scientists, advocates for those with degenerative disease, and politicians spoke and wrote of "the profound potential" of stem cells for medicine.[2] However, those who object to abortion, fetal tissue research and/or in vitro fertilization on moral grounds have condemned embryonic stem cell research and treatment in the strongest possible terms, advocating instead the use of stem cells derived either from adults or from blood obtained from the umbilical cord.[3] The scientific facts that would clarify whether adult-derived or embryo-derived stem cell therapy would be most efficacious are not yet in evidence, yet both pro- and anti-hES arguments in the clinical and bioethics literature have focused on science. There is thus much confusion about how the scientific facts of the matter relate to underlying moral concerns. Both sides have sought the middle ground, albeit without much success.

From the point of view of consumers, activists and patients, stem cell research may seem to have materialized from nowhere, a miraculous discovery with great potential.[4] Unlike contemporary genomics, which has become very much goal-directed and focused in character, the laboratories carrying out stem cell research had not just one or two therapeutic goals but in fact hundreds of possible research and clinical trajectories.

Moreover, stem cells have long figured prominently in basic research in human and veterinary cell biology, in clinical trials of possible therapeutic techniques, and even in a number of successful therapies.[5] Basic research involving stem cells is most often focused on fundamental problems of developmental biology; for example, how it is that specialized cells come into being, and how groups of specializing cells come to participate in coordinated activities.[6] Basic stem cell research thus focuses on the time in, manner through and extent to which somatic cells specialize during the development of an organism, and the role of stem cells in repopulation and repair of damaged or otherwise depleted cells in the mature organism.

Embryo Research as the Grounding for Stem Cell Research

For the purpose of this chapter, an embryo is the developing organism, understood to exist from the time of fertilization until the fetal stage. Human embryos became broadly available for research purposes only following the development of in vitro fertilization, developed in the 1970s by Steptoe and Edwards primarily to treat infertility. In 1978, Steptoe and Edwards documented the first birth through IVF; four years later, they reported their intention to freeze spare embryos for possible clinical or laboratory use. Since that time, scientists and clinicians have made use of embryos for solely research-directed purposes. At one level, it has been noted that embryos are the centerpieces of basic anatomy and pathology research concerning the basic units of a process of development. This research is also demonstrably useful in improvement of the clinical efficacy of in vitro fertilization, and for the investigation, at another level, of the diagnosis and treatment of hereditary and other diseases and injuries with the aid of pre-implantation genetic diagnosis (PGD).

The roots of stem cell research are to be found in understanding the chain of events and set of structures involved in processes of embryonic and fetal development. At the root of this interest is the question of how a human embryo transforms into a complex human being.

There are at least two kinds of hES cells; best classified are *totipotent* cells and *pluripotent* cells. The totipotent hES cells are found in the dividing fertilized egg. These cells have the unique ability to develop into any cell or tissue type found in the human body – for example liver, cardiac, nerve or blood cells – and in addition they have the capacity to form a complete organism. Pluripotent hES cells are found in the inner cell mass of the blastocyst: at the stage of development in which the dividing cell mass forms the shape of an almost hollow ball. While pluripotent human hES cells can develop into many if not all cell and tissue types, it is not currently believed that they would have the ability, if implanted in the human uterus, to divide and mature into an organism. Pluripotent stem cells are the cells most often used in embryonic stem cell research.

In order to obtain embryonic stem cells, the inner cell mass of a blastocyst must be isolated from its outer shell, removing the embryo from what would have developed into the placenta. Moreover the inner cell mass is disassembled by taking out individual embryonic stem cells for research purposes. The embryos used for hES cell research usually come from embryos created through in vitro fertilization (IVF) but not utilized for that purpose. The euphemism "spare" or "left over" embryo has been coined by clinicians and used by politicians to describe this source of cells for hES research and therapy.

Clinical Utility of Embryonic and Stem Cell Research

While it was not clear at the time of Gearhart's and Thomson's publications exactly what would result from the identification and cultivation of pluripotent hES cells, it was immediately apparent that their findings had great importance for both basic and clinical research in humans and animals:

- Gearhart and Thomson had identified a key point in the development of the human embryo at which the DNA in the nucleus of particular, undifferentiated cells no longer has the power to make another identical organism – the point at which totipotency is definitely not present.
- More importantly, these cell nuclei can produce a wide range, perhaps all, of the kinds of cells that populate a developing or mature human organism.
- It is possible to derive these cells from the embryo and to isolate them from other cells.
- Once derived, these isolated pluripotent human embryonic stem cells can be cultured and frozen, transported and grown, fed and measured in a variety of ways.
- These cells can be induced to produce differentiated cells. Thereafter, these cells might themselves produce more cells, which might be

> transferred from culture into the bodies of patients to replace a wide variety of damaged cells or to perform a range of other tasks, from inoculation to the destruction of cancerous tissues to the delivery of drugs.

Several well-publicized clinical trials involving the transplantation of fetal tissue into patients with degenerative diseases of the brain and nervous system, such as Parkinson's disease, have been conducted with essentially no success, despite successes using essentially the same modality in mouse and primate trials. While these trials did not specifically measure the activity of stem cells, they raised basic questions about the utility and toxicity of immature cells for transplantation. Clinical research that specifically involves stem cells has included a wide variety of tests regarding the effectiveness of transplanted stem cells in repopulating certain needed cell types in patients suffering from bone cancer and diseases of the immune system.[7] Techniques already in use include the harvesting of stem cells from umbilical cord blood, and the transplantation of stem cells for the treatment of leukemia.

Enthusiasm about embryonic stem cell research quickly led to a larger discussion of the future of the work, and the implications of stem cells for broader debates about how to allocate healthcare resources, how to proceed with caution in new areas of clinical research, and how to regulate research involving embryos, fetuses or abortion. Wide calls for government investment in stem cell research were entertained both as part of the 2000 presidential campaign in the United States and as part of governmental hearings. It has been noted in the United States and also in the course of the ECSSR conference discussions that, like mammalian cloning research, researchers whose work was funded by small companies rather than national or regional governments were achieving greater innovation in stem cell research. Arguments for government funding of stem cell research were almost always linked to the claim that the government funding would enable regulation, and if necessary restriction, on stem cell research. This argument received the endorsement of many ethics advisory boards, including, for example, the US National Bioethics Advisory Commission (NBAC), an arguably partisan board of ethicists appointed by President Clinton.[8] What did not emerge immediately was the question of how patents filed in association with Gearhart, Thomson and others might make it difficult for the government to exercise as much regulatory authority as groups like the NBAC, the American Association for the Advancement of Science (AAAS) and others had sought.

Ethical Issues in Contemporary Research and Therapeutics

While the subject of research on embryos presents a variety of ethical and legal issues, the central issue in the Western debate has long been the moral status of the embryo. This debate is not unique to twentieth and twenty-first-century scholarship in science and bioethics. On the contrary, this controversy originates and is deeply rooted in different religious and philosophical views. In the Western philosophical tradition, the debate over the status of the embryo can be traced back to Aristotle, who wrote of the ensoulment of the human at a particular stage, as did the pre-Socratic philosopher Heraclitus before him. Religious views of conception have been extensively debated in Judeo-Christian and Muslim scholarship dating to the earliest religious texts in those traditions. The contemporary question of the moral status of the embryo emerged during the US controversy over the legality of abortion in the 1960s–1980s, and continues to be an issue in the discussion of the use of most reproductive technologies today. Views on the moral status of the human embryo normally take one of the following three forms:

1. The human embryo has no intrinsic moral status; it derives its value from others.
2. The human embryo has intrinsic moral status, independent of how others value it.
3. Embryos begin with little or no moral status and continue to achieve more and more status as they develop.

Position 1

The position that an embryo has no moral status can be argued in different ways. Since the fetus depends completely on the pregnant woman for development, many ethicists believe that it cannot be viewed as a unique entity. Instead several, including the famous MIT philosopher

Judith Jarvis Thomson, argue that the best metaphor to describe the status of the fetus is that of a parasite (whether desirable or not), possessing no moral status independent of the mother. Those who hold this position do not object to embryo or fetal research on the grounds of the moral status of the fetus, and would refer, for example, to fetal surgery (whether conducted *ex utero* or *in utero*) as a procedure, strictly speaking, on the mother. The concerns expressed by those who hold this position about embryo research are focused on the long-term social implications of embryo research for the status of born persons, particularly those with disabilities. However, it is not held that the destruction of an embryo is inherently morally problematic.

Position 2

The position that the fetus has intrinsic moral status is grounded in the view that a *person* is created at a moment in time that can be linked both to the consummation of an act by those who participate in its creation, and to the physical and legal initiation of that person's participation in the human community. The metaphor most often used to describe the status of the fetus for these purposes is that of *baby*; the ever-increasing presence of the fetus in public and private life has contributed to the view that from the "moment" of conception a person can be identified, independent of the risks that face a person so defined and regardless of the plain differences between such a person (for example in the case of a frozen embryo) and a person who participates as a baby, child or adult in the institutional life of the community. Given this view of conception and the embryo, the use of an embryo for research purposes is exactly tantamount to the use of any other vulnerable subject in research without consent; research that poses not only a great risk but in many cases has the clearly anticipatable outcome of death for the subject.

Position 3

Several philosophers and scientists have argued for a developmental model of the moral and legal status of the human embryo and fetus. This position begins with the claim that clinical changes in the embryo and fetus have moral significance because they represent, if not ensoulment, the development of concomitant ability of the being to participate in the human community. One way in which this position has been expressed is in the Roe decision, which held that pregnancy can be divided into three periods, corresponding to the degree to which the embryo has developed, and the opinion issued in that case by the Supreme Court to the effect that these periods represent the increasing standing of the emerging human person in the human community. Contemporary neonatal technology has made it possible to construct a clinical definition of *viability*, a time at which the developing fetus would be able to survive outside the womb. Important, though, is the fact that not only does the fetus change over the course of pregnancy, but the technologies of neonatal care evolve as well, so that in the course of five years the moral status of a twenty-two-week fetus would change with the state of the technology, rather than remain fixed at some natural point in development.

Those who hold that the development and viability of an embryo is morally relevant to research on embryos, fetuses and stem cells must face an interesting array of problems: how can values (e.g. the rights of the embryo) be derived from facts? What moral status is conveyed to a laboratory creation such as an embryo-like creature made from parts taken from several different species? What, if any, moral standing does the specialized cell in an adult have if it can be demonstrated that all that is required to turn that adult cell into a cloned embryo is a jolt of electricity or bath of enzyme? Polls indicate that the above position, held by the majority of registered voters in the United States and United Kingdom, is in many ways the most complex in virtue of its attempt to be responsive to changing science.

Law and the Ethics of Embryo Research

The moral issues surrounding embryo research leave the status of the embryo highly contested. The difficulty for the law in dealing with this terrain, is that it is fraught with confusion and presented to the courts in the form of a particular case – and often in a context where little expertise is available or admissible on the subjects of the science and ethics of the matters at hand. Moreover, in courts, as opposed to the institutions of religion and philosophy, some consensus must be reached in order for the institution to complete its appointed task in each case.

Even when the Supreme Court in the United States passes over a case on abortion, it has taken an action that is important both for the purposes of allowing others to determine where the law stands and for the purpose of the completion of the judicial task. This is inherently problematic for a number of reasons: not only does relatively little agreement exist between scientists, ethicists, lawyers, lawmakers and religious leaders in regard to the status of the embryo, but it is also unclear what path embryo research should take from the perspective of the "experts about the matter" because, in a legislative, judicial and economic leadership vacuum, it is difficult to determine who the experts are.

The lack of consensus about the status of the embryo and the morality of research has resulted in what might be somewhat contradictory and unclear legal definitions in the United States at the state and federal level. Since it is extremely difficult to define the status of the embryo, the question still remaining hotly contested, most of the legislation tries to steer away from making a definitive statement that would outrage either side of the debate. The legality of embryo research also varies from country to country.

Experimentation on the embryo for the purposes of developing stem cell and other technologies, and for general knowledge, is legal in the United Kingdom and three Australian states under certain circumstances. In Germany, embryo research is banned completely. In the United States, debates over the legality of embryo research tend to hinge on prior state court holdings, federal agency rules and directives, or state laws on the status of the embryo.

Even though the courts already attempt to resolve the debate over the status of the embryo, they must also undertake a new set of questions: should experimentation be allowed at all? If so, under what circumstances should it be prohibited? For the majority of US residents, at least, polls show that some experimentation is desirable, so the question the courts face in the political arena is where the line between acceptable and unacceptable experimentation should be drawn.

Law and the Status of the Embryo

Historically and under common law, the fetus has not been legally protected until after complete separation from the mother's body. This view holds that because the fetus is not independent *in utero*, it cannot possess individual rights. Consequently, the fetus *in utero* has not been legally protected against any harm. Recent decisions criminalizing the termination of pregnancy or even activities that might result in eventual harm to a potential future person under certain circumstances have altered the tradition of common law concerning the fetus and embryo, as have lawsuits concerning wrongful birth. Specifically, mother and child are now able to make a tort claim for malpractice that takes the form of medical negligence if predictable harm to the embryo *in utero* has had a negative effect on the newborn child.

For purposes of defining the status of the embryo, courts have also relied upon the *personhood test*: when, and under what circumstances, and given what kind of creature, is an embryo considered an embryo, and when is it considered a person for the purpose of legal protection? In the case of Roe v. Wade (1973), the United States Supreme Court denied that the unborn be considered "persons" under the Fourteenth Amendment. However, they failed to set forth a clear definition of personhood or to explain why they denied the unborn such status. Consequently, there has still been much debate concerning the legal status of embryos, even given the aforementioned court interest in recognizing three periods of pregnancy and the clinical and legal significance of the third period or trimester for assigning increasing interests (if not rights) to the fetus.

Since Roe v. Wade did not clearly define personhood, the Court had to use other means to construct a definition of an embryo. This task, like most others involving embryo experimentation, was and remains highly problematic. It is a task that has been taken up in many nations and states, one contingent on whether fertilization should be assumed to confer individuality and, if so, if fertilization is an event or a process. It is as a result of debate on this matter that the courts and their advisory bodies, and legislation, have come to focus on the metaphysical question of identity, and whether or not personhood or individual identity ought to play a part in determining at what point an embryo is too mature (and thus possessed of moral standing) to be subjected to involuntary (and thus, any) testing.

The Warnock Committee published a report in 1984 stating that destructive embryo research should only be permitted up to 14 days into development. The 14-day limit was based on the following arguments:

- Twinning can occur up until 14 days of development.
- If twinning is still possible, then an embryo cannot be considered an individual.
- Only individuals can have moral status.
- Beings without moral status have no right to be free from destruction and thus can be subjected to experimentation.

The 14-day rule rests on the assumption that being an individual confers moral status on a being, and provides its own definition as to when this individuation occurs. However, other standards have also been proposed. One is the constantly evolving notion of viability: perhaps the viable fetus has moral standing, while the fetus that cannot survive outside the womb does not. Another is the standard of birth or even of informed consent with parental surrogacy, which would either rule in or rule out embryo and fetal research depending on one's (or one court's) view of the importance of informed consent or the nature of surrogacy for a fetus. This question has been raised in fetal surgery. Still another is the assertion that at conception or fertilization there is a person in place, but here the question remains at what moment the actual fertilization or conception takes place, and under what circumstances one could perform any clinical or research procedures on a *conceptus*, and by whose authority.

Foundations of Ethical Debate in Stem Cell Research and Therapy

It has been maintained that one's position on the ethics of stem cell research depends not only on the question of when conception occurs and what bearing each developmental milestone has on the moral standing of a fetus, but also on the underlying view one holds about values and ethics; one's ethics will determine the horizon of the moral inquiry, one's view about whether a moral matter tends to involve personal choices by involved actors who are rational, or is instead a broader and more social dialogue leading to either a social contract or the creation of social institutions, will bear on whether one is willing or capable of engaging in deliberative democratic discourse on this complex set of questions. There are several theoretical questions in ethics that are of this variety.

Theory of Rights

It is claimed by some that because the embryo has no "interest" in living, it does not possess any right to live. This argument rests on the assumption that killing is wrong because it deprives a person with an interest in life of his/her necessary interest in life. If an embryo is neither conscious of life nor cares for the duration of its own, it has no intrinsic moral status under the theory of rights articulated by Robert Nozick and others. Specifically, it has neither a *positive* right to be thawed out from a nitrogen tank and given a womb, nor a *negative* right against being demolished while proceeding through development ensconced in the womb. The emphasis is on liberty interests attached to the idea that a person is rational and capable of articulating interests, an emphasis with a number of weaknesses and strengths when elevated to a legal and moral argument.

Consequentialist Theory

For the consequentialist, an action's moral status is determined by the ends it serves, and good ends justify the means necessary to achieve those ends. Embryos can be experimented on or even destroyed, consequentialists have argued, because the ends of embryo research outweigh whatever damage is done – including the destruction of embryos – as long as it is clear that the embryo's suffering or death is not more morally undesirable (to itself or to others, understood in a variety of ways) than is the suffering of the patient or community or family affected by a treatable or potentially treatable disease under investigation, the treatment of which uses stem cells that require the destruction of embryos.

Religious Views

A number of religions express views about abortion and indeed about reproduction and research that have been debated in intra-denominational and social forums. It is important to take note of the view held by the Vatican since 1859, because that view is in play in the political debate more than any other in the West. This is the view that the embryo obtains moral status at the moment of fertilization. Recently, the Vatican has gone so far as to link fertilization and moral standing to genes – with a unique genetic make-up, an embryo is given a soul. Since twinning can occur up until the fourteenth day of development, and two zygotes can fuse, a theory of individual ensoulment predicated on genes and fertilization faces scientific hurdles no less than other views.

The Derivation Dilemma

Whatever its religious or scientific underpinnings, the ethical debate surrounding hES cells has recently centered on how the hES cells are derived and on whether or not they should be protected from destruction, much like an adult is. Using leftover IVF embryos for the purposes of hES cell research raises complex questions about the status of the embryo, the value of human life, and whether there should be set limits regarding the interventions into human cells and tissues. Furthermore, questions about adequate informed consent, oversight and regulation also come prominently into play.

Those who support hES cell research argue that an embryonic stem cell, even though it is derived from an embryo, is not itself an embryo and thereby would never continue to develop into a fetus, child and adult. Each stem cell is only a cell that can be triggered to become a specific kind of tissue yet could not be triggered to become an individual. Furthermore, the embryo at the blastocyst stage has not developed any kind of nervous tissue and thus extracting individual stem cells would not be painful for the embryo. Since the embryos used for stem cell research come mostly from the leftover IVF embryos, which would otherwise be discarded, the proponents of stem cell research argue that it is better to use such embryos to find cures for debilitating diseases rather than to discard them, benefiting no one.

It is also argued by many that the embryos used to make hES cells are not embryos at all but instead something else, either "pre-embryos" or cells that are only partially human. In many cases, no conception occurs in the creation of these cells; for example, in the case of nuclear transfer to make a genetically identical embryo-like human that grows into a blastocyst but might not be able to survive implantation in a womb. What is an embryo, and what does it mean to make something that behaves like an embryo but could not come to full term in a womb?

One attempt to resolve the debate over stem cell research involved the suggestion that researchers might obtain stem cells from embryos without actually engaging in the destruction of those embryos.[9] This was originally proposed by the US National Institutes of Health under the Clinton administration, and was in substance taken up by President George Bush, who suggested that while it is in his view immoral to destroy embryos, some hES cells have already been derived from embryos that have already been destroyed – and the matter of the availability of those cells can be considered distinct from the matter of creating new cells through the destruction of additional embryos. He thus decreed that only stem cells derived from embryos destroyed prior to his speech would be made available for federal funding. As the President framed his compromise, "only those cells for which the life or death decision has already been made" would be eligible for use.[10] He noted that 66 stem cell lines have already been obtained from embryos, "more than enough" to allow that research to proceed.

Predictably, a number of concerns were raised about the President's rationale and his policy. However, the overriding question was whether enough embryonic stem cells did in fact exist. The issue of the suitability and scarcity of hES cell lines already derived at the time of Mr Bush's speech in turn called attention to the fact that many human embryonic stem cell lines are subject to US and international patents, and that many of the innovations necessary to derive, culture, differentiate or otherwise manipulate stem cells are also subject to patents.[11] Yet, *should* stem cells, embryos, embryo-like organisms or the cells derived from them be eligible for consideration as intellectual property, whether through patents or other protections of law? Is the compromise made by President Bush in fact a compromise of the principle that life begins at conception, and thus a political attempt at consensus, or merely a way to attend to the problem of the political realities of overwhelming support for the research, set against an incredibly vocal minority opposition – who constitute the bulk of the conservative party for purposes of the debate about abortion.

Another central problem is the permissibility of making embryos specifically for research purposes. There are two different types of embryos used: those classified as "spare" embryos which are left over from unsuccessful in vitro fertilization and those cultivated specifically for purposes of being tested. Some people have ethical concerns about both of these methods; however, those who support research are more likely to question the ethical nature of the second of these two alternatives.

The argument that it is acceptable to use spare embryos but not to create embryos specifically for that purpose centers on Kant's categorical imperative, specifically the formulation of that imperative that centers on the claim that the ultimate moral wrong is to treat someone as a means to some other end, rather than as an end in him- or herself. Those who do not support the use of embryos for the sole purpose of enhancing research argue that it is morally unacceptable to use embryos for scientific purposes on the grounds that this is a clear use of a person as a means. Some of these same arguments can apply to the use of embryos under any circumstances. In the case of spare embryos, by contrast, many are too old or morphologically inappropriate to be implanted, and thus have no other use. It is thus argued that the use of these for research is not nearly as questionable. Moreover, opponents claim that if the cultivation of spare embryos is legalized, scientists will act on the incentive to produce as many embryos as one could produce. Even many of those who do not oppose the creation of embryos for research on Kantian grounds have voiced concern that creating embryos merely for research might cheapen the act of creation.

Clinical Implications for ART Clinics

Whatever the form of embryonic stem cells to be utilized in research, the involvement of clinical assisted reproductive technology (ART) embryologists, technicians and clinicians is omnipresent. The processes whereby embryos are created, analyzed, stored, removed from nitrogen freezing, or destroyed are all processes that require the technologies, clinical expertise, patient population and institutions of assisted reproductive technology. This is true whether such embryos are created from donor eggs and/or sperm intended for research purposes, or as a by-product of reproductive healthcare. It is thus no surprise that the largest research programs to date in the field have employed obstetricians, andrologists, reproductive endocrinologists, and even ART psychologists and social workers.

Ethical issues involved in participation in stem cell research includes three key issues:

- Whether, and under what circumstances, patients or research subjects should be allowed to participate in the donation of reproductive materials for stem cell research, particularly where that research involves the creation of embryos for research purposes.
- Whether reproductive clinicians and technologists should be involved in the non-reproductive use of cloning technologies for the creation of nuclear transfer-derived stem cells.
- Whether and when clinicians involved in the derivation of embryonic stem cells should be held responsible for the failure of those cells in clinical trials or therapies using those cells.

There is no professional consensus at this point on any of these issues, although all three will receive the attention of bioethicists as well as ethics boards of professional societies such as the American Society for Reproductive Medicine (ASRM) in the United States.

6

Bio-Terrorism and National Security

*Sue Bailey**

In this new millennium, there is a burgeoning threat to world peace and security. It is as old as time but represents a growing threat to the nations of the earth. The international community is now faced with weapons of mass destruction on a level heretofore unimagined. The balance between war and peace has been altered in scale and scope by scientific advances that provide disastrous capabilities to the perpetrator, be they combatants or terrorists.

Bio-terrorism is one of the most frightening weapons in this new age of terrorism. It has been estimated that anthrax, for instance, can kill as many people as a nuclear detonation. Indeed biological weapons are often described as the poor man's atom bomb. Furthermore, smallpox because of its lethality and contagiousness is a threat of epic proportions on a worldwide scale. Weapons of mass destruction are not restricted to biological or chemical agents but also include nuclear weapons and the continuing threat of conventional weaponry.

During the year 2002, the United States experienced terrorism from an unknown source in the form of anthrax dispersed in powder form through the mail. The resultant mortality and morbidity, though limited in scope, drastically affected the country, the government and its people. As disruptive, disturbing and deadly as the attacks were, they pale in comparison to mathematical estimates by the World Health Organization

* Dr. Sue Bailey is a former US Assistant Secretary of Defense.

that under certain conditions, anthrax delivered by other methods could result in casualties on a scale similar to those caused by a nuclear weapon.

Biological Agents

There is a range of biological agents that can be used in bio-terrorism but fortunately many are unstable, volatile and difficult to obtain, produce or weaponize. The list is extensive but those most often included as the most serious bio-warfare agents are anthrax, smallpox, pneumonic plague, botulism, tularemia and hemorrhagic fevers.

Anthrax

During medieval times, around 80 AD, the Huns described an illness that killed their horses and cattle. In the Old Testament, a disease described as the "fifth plague" killed many humans. Anthrax was the likely cause of these early epidemics. In 1770, approximately 15,000 people died of intestinal anthrax in Santo Domingo. Anthrax outbreaks still occur in various parts of the world, but the actual number of human victims infected today is relatively small. These incidents typically involve farmers, veterinarians, and mill workers who work with infected animals or animal products; cases such as these are usually documented as having contracted "wool sorter's disease." This name and others actually refer to the diseases caused by the anthrax bacterium, *Bacillus anthracis.*

Anthrax is an ancient, naturally occurring bacterium that has been re-engineered and 'weaponized' in modern times for bio-warfare applications. Today, there are many different strains, or unique DNA variations, of the bacteria. When viewed through a microscope, this bacterium has a distinctive rod-like shape and linear organization that resembles the boxcars of a train. Anthrax can exist in a dormant spore form for long periods of time. These spores are resistant to heat, sunlight and even many disinfectants, making anthrax a prime choice for terrorists who seek a biological weapon. Once in contact with a human or other host body, the spores may germinate into live bacteria.

Like many other germs, anthrax can coexist with humans and may not necessarily cause an infection. The human body is, in fact, a host to numerous types of bacteria that are generally harmless unless their populations reach unusually large numbers or take up residence in areas of the body that may be vulnerable. In cases where humans have come into contact with anthrax, the number of spores present largely determines whether or not an infection will occur. Other factors such as age and the strength of an individual's immune system may also determine the likelihood of infection and the severity of the resulting disease. The manifestation of the disease (e.g. the type of illness contracted) depends on the way in which the bacteria entered the host's body.

For instance, gastrointestinal anthrax may be contracted by consuming undercooked meat. Once ingested, the spores may germinate in the abdominal or the oral-pharyngeal regions of the gastrointestinal tract. The abdominal form affects the lower gastrointestinal tract with symptoms that include nausea, vomiting and diarrhea. Sepsis can also occur when toxins released by the bacteria get into the bloodstream. The oral-pharyngeal variation affects the mouth and throat or upper gastrointestinal tract where lesions or sores form and cause swelling. The lymph nodes become enlarged and this can also lead to sepsis.

If spores entered the body through a wound or even a scratch on a host's skin, cutaneous anthrax may occur. The skin lesions themselves are generally painless, but the associated swollen lymph nodes may be painful. Reported anthrax lesions are usually seen on exposed skin such as the face, neck, arms and head. They are a result of direct contact with contaminated materials but can also occur following an aerosolized release of spores that then settle on the skin. The incubation period following initial exposure to anthrax spores ranges from one to twelve days. The first symptom is usually a small lesion resembling a bug bite that soon evolves into an ulcer or open crater-like sore. A characteristic black scab subsequently forms on each lesion. The color of the scab yields the disease's name, which comes from *anthracite*, the Greek word for coal. During the acute phase of the illness, the site and nearby lymph nodes may become inflamed. The scabs usually fall off in about two weeks leaving little or no scar. There can also be systemic signs of the disease like fever, headache and malaise, or lack of energy. There were 224 cases of cutaneous anthrax reported in the US between 1944 and 1994. Recently, there have been about 2,000 cases reported worldwide per year. With antibiotics, progression of cutaneous anthrax to systemic disease may be virtually eliminated, although antibiotic therapy does not usually inhibit the progression or healing of the skin lesions. Without therapy, the cutaneous infection may transform into systemic disease and the resulting death rate may reach 20 or 30 percent.

The deadliest manifestation is known as "inhalation anthrax" or "pulmonary anthrax" as it occurs following inhalation of anthrax spores. This type of infection occurs only when a large number of spores are inhaled into the lungs. Medical scientists are uncertain about the number of spores that must be present in order for disease to occur. Many experts believed that approximately 10,000 spores must be present in order for subsequent disease to occur, but recent clinical observations and research data indicate that the actual number of spores needed to infect may vary rather widely from person to person. This variability may be explained by the fact that some people are more susceptible to or less able to fight off inhalation anthrax and could be infected with far fewer spores. There is still too little data from which to draw conclusions about susceptibility, but it seems possible that there may not be a magic number of spores required for disease to occur after all.

Anthrax was first used as a bio-warfare weapon during the First World War when reindeer were used to haul supplies to the allied troops. Reportedly, there was a plot to infect the reindeer with anthrax. A German was found to have vials of anthrax and it was thought that he planned to use it to infect the animals. When the contents of these vials were tested in 1998, the spores were still viable!

During the Second World War, virtually all countries involved in the war had a bio-warfare development program and most of these programs included the weaponization of anthrax. In 1942, Great Britain experimented with anthrax on an island off the coast of Scotland. They set fire to the island following the experiment in an attempt to destroy the anthrax, but an alarmingly high spore count persisted even after the fire had subsided. In 1943, the United States started its own research involving biological agents such as anthrax. Related projects continued until President Nixon issued an executive order in 1969 that halted the program. The stockpiles were destroyed in 1971 and 1972.

In Sverdlovsk, in the former Soviet Union, an accidental release of anthrax from a military research facility known as Compound 19 resulted in the worst outbreak of anthrax that has been documented in recent history. People living downwind from Compound 19 developed fever and had difficulty breathing. Ultimately, the death toll was estimated to be between 200 and 1,000. Originally, the release of anthrax was not reported and the disease was blamed on contaminated meat. Years later, in 1992, President Boris Yeltsin acknowledged that the outbreak was indeed caused by the accidental release of anthrax spores.

Following the Gulf War in 1991, United Nations inspection teams verified Iraq's bio-warfare capabilities and research that explored the use of biological weapons, including anthrax. It was the first acknowledgement by any government in recent times regarding plans to use biological weapons. UN teams discovered that bombs had been loaded with the pathogen and that Iraq had in fact produced 8,500 liters of concentrated anthrax (6,500 liters had been placed into munitions). The validation of the UN teams' intelligence reports prompted the inoculation of US troops against anthrax.

It is not known what current capabilities and stockpiles remain in the hands of other countries but concern continues to grow as the world finds itself confronted with the reality of bio-warfare.

The United States is actively preparing to deal with the risk of terrorism and bio-warfare. Doctors and healthcare professionals are rapidly increasing their knowledge of and ability to manage the medical consequences stemming from weapons of mass destruction. The public health system and local and federal authorities are setting new requirements for bio-defense. First responders at the local and state level are coordinating their efforts and now have the ability to respond more effectively through detection, surveillance, treatment and sophisticated communication.

Smallpox

Today, the world struggles with the pressing concern that the smallpox virus, once thought to be confined to two secure sites, could, in fact, be in the hands of those who would once again unleash its deadly fury on the earth and its inhabitants. In fact, there has been deliberation among world leaders and their top scientific advisors, who have debated destroying what was thought to be the last smallpox cultures. They have been retained for research purposes, out of concern that new information that can only be obtained from study of the original culture might be necessary in the future. If ever there were any cases or outbreaks of smallpox, it would signal an intentional release of the original virus.

Smallpox has plagued man for many centuries, wreaking havoc, sometimes even destroying cultures and disabling armies. The Aztecs, for instance, were infected by the Spaniards who spread the disease among a population with no defenses. In fact, the civilization was largely destroyed by the virus.

In the twentieth century it is estimated that as many as three hundred million people died from the disease. Prior to the advent of the smallpox vaccination, there was no form of prevention and even today there is no real treatment.

The World Health Organization declared smallpox eradicated in 1980 but, as outlined, there is growing concern that the remaining viral samples are no longer secure and may have fallen into the hands of those who would use it to wage war or terrorize enemy targets or populations. Vaccinations essentially ceased in 1980, leaving the world unprotected from this deadly disease.

It is no wonder that as a species we are so concerned about the possibility of smallpox re-emerging in the form of bio-terrorism or bio-warfare, and there are some historical precedents for this. The British may have attempted to infect Native Americans by providing blankets previously used by smallpox victims. It is also known that Japan considered using it as a weapon in the Second World War.

Most recently, the United States and other countries have undertaken plans to re-institute a vaccination program for the general population if necessary. This is a clear indication of the level of concern about the possibility that smallpox could be used for bio-terrorism.

Natural outbreaks of smallpox prior to 1980 were controlled through what is termed "ring vaccinations." Ring vaccinations include the immediate inoculation of those affected and their contacts. It was highly successful in very small communities but clearly will be far less effective or even useless in a highly populated, mobile society.

Researchers at Yale's School of Management and School of Medicine reported in the *Proceedings of the National Academy of Sciences* (NAS) in July 2002 that mass vaccinations would be more effective and save more lives than ring vaccinations. It was concluded that 560 lives would be lost to smallpox if mass vaccination were instituted as opposed to 4,680 with the more limited plan. The study was based upon an outbreak in a city of 10 million people.

Smallpox is an orthopox virus that occurs as either the variola major or the variola minor strain. Poxviruses are a group of DNA-containing viruses with a common antigen. They are identifiable from each other depending upon the severity of the skin lesions they produce.

Smallpox is a contagious disease, though not as contagious as measles or influenza. It is transmitted through direct contact or through droplets in the air but generally requires face-to-face contact for transmission. In the early stages, the disease is transmitted in nasopharyngeal secretions but eventually the lesions are themselves infectious. The incubation period is about nine to fifteen days. The individual is most infectious in the first week of the disease but can still be contagious until the lesions scab over and the scab falls off.

The disease goes through several phases, including the prodromal phase, the early eruptive phase and then the vesiculation and pustule phase. It begins with flu-like symptoms that can include fever, headache and muscle aches. After three or four days, the fever subsides and it is at this time that the characteristic skin lesions begin to form. They usually begin in the mouth and on the face and then move to the extremities and the trunk. They can remain as scattered lesions or begin to cover most of the body. This disease progression is classic for smallpox in its moderate form but a much more deadly form can involve hemorrhagic lesions, bleeding from the mucosal membranes, followed by shock, coma and death within three or four days following the incubation period. There is no real treatment other than supportive care. There is some hope that antiviral drugs could be helpful.

Toxins

Toxins are another great bio-terrorism concern and are generally described as poisons or toxic substances produced by plants and animals. Fortunately toxins do not reproduce like a bacteria or a virus. They are not considered volatile like chemical agents and do not normally affect the skin. Most would therefore have to be delivered as an aerosol and are difficult to produce in large quantities. However, because of their lethality, they are considered a significant bio-warfare threat.

Toxins that are of high concern are staphylococcus enterotoxin B, ricin, T-2 mycotoxins and botulinum toxin. Botulinum is considered to be one of the deadliest substances on earth and is of particular concern. There is a pentavalent toxoid for botulinum but the main treatment is decontamination and ventilatory support. Symptoms include slurred speech and weakness that progresses to paralysis.

Chemical Agents

The use of chemical agents in warfare goes back centuries and was documented during the Peloponnesian War when sulfur and coal smoke were used as a weapon and blown through a hollow log. That war pitted the allies of Sparta against the Athenians. The Greeks added pitch, naphtha, lime and saltpeter to the smoke and it was called Greek fire.

With advances in chemistry in the eighteenth century, the interest in chemical warfare grew, eventually stirring the debate that rages today over the associated moral and ethical issues. Many nations have engaged in the ongoing struggle between scientific molecular capability and the rules of engagement and the ethics of war. A resolution against the use of chemicals on the battlefield was included in the Hague Convention of 1907.

In the First World War, Germany and Britain used chlorine and phosgene in the theater of battle, but in 1917 the German shells also contained sulfur mustard. The latter not only created pulmonary symptoms but was also persistent and stable, and contaminated soldiers who touched it as well as breathed it in. These agents changed battlefield uniforms, which from then on necessarily included gas masks and, eventually, protective suits. Though fewer than 5 percent of the casualties died, the effects and recovery went on for weeks, overwhelming the medical support system. It was a clear demonstration of the effectiveness of a chemical agent in war, a dreadful harbinger of emerging trends in terrorism.

During the Second World War it is now known that Germany had developed Tabun, a toxic organophosphate, and Sarin, a similar but even deadlier nerve agent. It was not used on the battlefield, but cyanide, a lethal chemical agent, was used in the concentration camps. It is possible that Japan used chemical agents against China but other than that inconclusive report there was no known use of chemical warfare during the Second World War.

Since the Second World War, research and stockpiling of chemical weapons has continued, as have reports of usage in battle and other conflicts. The United States used defoliants and riot-control agents in

Vietnam; Egypt was accused of using mustard against North Yemen; and the Soviet Union was said to have used chemicals against Afghanistan. In the 1980s Iraq was said to have used chemical agents against Iran, and a United Nations investigation confirmed the use of mustard and Tabun. Later it was reported that Sarin was used by Iraq. Iran may have retaliated with chemical weapons as well.

Since that time, chemical agents have continued to be studied, stockpiled and traded around the world with little oversight on a national or international level. At the time of the anthrax attack in the United States there were as many as 250 "culture centers" selling, for a small price, cultures of deadly biological agents. Security was often the responsibility of the public, academic or private organizations themselves. Today as many as 24 nations have the ability to produce chemical weapons. Although the United States has some chemical weapons they are under a congressional mandate to destroy them. Incinerators are now in operation, with others being planned, to destroy the existing chemicals.

There are a variety of chemical agents for offensive use, whether offensive warfare or terrorism. However, their usefulness for these purposes is dependent on many factors, including volatility and persistence. They can be effective in several different forms including solids, liquids, gases or vapors. The form is related to temperature and pressure. Given these parameters, the effectiveness of any particular agent relates to weather, wind conditions and the environment in which exposure occurs.

Pulmonary Agents

Pulmonary agents have their primary effect on the lungs, causing shortness of breath and edema or excess fluid in the lungs. It can also produce eye and general airway irritation. Decontamination is accomplished mainly with large amounts of water for liquid contamination, and by escape to unpolluted air or the administration of oxygen for vapor exposure.

The First-World-War chemical weapon phosgene is the principal pulmonary agent that has been used offensively. Other known pulmonary hazards that could be used as a terrorist agent or warfare hazard to the lungs are perfluoroisobutylene (PFIB), zinc smoke and oxides of nitrogen.

Cyanide is considered a pulmonary warfare agent but is only lethal in very high doses that are difficult to maintain in most conditions. This category of agents also includes hydrocyanic acid (AC) and cyanogen chloride (CK). Exposure to cyanide results in seizures, as well as respiratory and cardiac arrest. Because of the quick evaporation and dispersal of cyanide, decontamination consists mostly of removal of clothing and administration of water. Intravenous sodium nitrite and sodium thiosulfate are the antidotes but management also requires the correction of acidosis and the administration of oxygen.

Vesicants

Vesicants or blister agents include sulfur mustard (H, HD), lewisite (L) and phosgene oxime (CX). The vesicles or blisters produced generally affect the skin but also the eyes and airways. Vesicants all act in a fairly similar manner. They cause skin blisters, pain and irritation to the eyes and airway. Sulfur mustard has a latent period of several hours while the others cause immediate irritation with progressive effects. Sulfur mustard can also cause gastrointestinal effects and bone marrow suppression.

Decontamination for all blister agents is done with 0.5 percent hypochlorite and water and is essential to stop further damage. British anti-lewisite (BAL) is an antidote for lewisite. There is no treatment other than supportive care.

Incapacitating Agents

Incapacitating chemicals include BZ and another chemical Agent 15, which is probably closely related to BZ. During the Moscow theater hostage crisis in October 2002, a general anesthetic (fentanyl) was apparently used as an incapacitating agent against the terrorists and demonstrated how anesthetics, soporifics or sedatives could be deployed in this manner. Riot control agents such as CS (a form of tear gas) and CN (mace) have also been used to incapacitate. Symptoms of incapacitating agents depend upon the agent. Where irritants are involved, fresh air and water decontamination are recommended. In the case of a narcotic agent, the antidote narcan and immediate supportive medical and ventilatory care are required.

In 1997, 147 nations ratified the Chemical Weapons Convention (CWC). The accord outlines the use of nerve gas, including, for instance, mustard gas, VX and Sarin as well as other lethal chemical agents. It does allow less dangerous "law enforcement" substances such as tear gas, pepper spray and mace. Other incapacitating agents like fentanyl are not allowed for military purposes but are permitted in carefully controlled domestic situations.

Nerve Agents

Finally, nerve agents are a major terrorism concern as they are among the most deadly of the major chemical agents. Those agents causing greatest concern at this time are GA (Tabun), GB (Sarin), GD (Soman), GF and VX. They can kill within minutes by inhibiting acetylcholinesterase, an enzyme that prevents excess amounts of a neurotransmitter, acetylcholine, from altering the normal functioning of the nervous system. They cause respiratory distress and excessive secretions in small doses of the agent in vapor form but with increasing exposure can cause loss of consciousness, convulsions, paralysis and death. In the liquid form, small doses result in sweating, weakness, nausea and vomiting. Heavy exposure to the liquid form causes symptoms similar to and as dangerous as the vapor form.

Defense against nerve agents includes detection systems to indicate the presence of a nerve gas and protective clothing, including a gas mask. Immediate treatment depends upon the particular agent but generally includes atropine, praladoxime and diazepam, as needed, according to the level of exposure. Decontamination and supportive treatment are essential.

The Sarin attack in the Tokyo subway system in 1995 clearly demonstrated the difficulties in dealing with a nerve gas terrorist attack. In that case, detection was based upon symptomatology. Decontamination was insufficient, antidotes relatively unavailable or administered too late, and the hospitals overwhelmed with casualties. This nerve gas attack is an example of the difficulties encountered in asking first responders to cope with a chemical agent rarely seen away from a battlefield. Furthermore, it indicates that the speed with which even a sophisticated medical system acts can be insufficient for dealing with mass casualties. In many ways, military combatants on the battlefield may be more at risk but are generally better prepared than civilian victims of chemical terrorism.

Conventional Weapons

Terrorists often use conventional weapons and simple means of deployment when carrying out terrorist activities. Bombs and other explosive devices are the most widely used conventional weapons. Guns, mines, hand grenades and rocket-propelled grenades are also part of many terrorist arsenals. The use of missiles is rare but a few groups are known to be in possession of surface-to-air shoulder-fired missiles capable of bringing down helicopters, military aircraft and civilian airliners. While large numbers of powerful guns are manufactured and are readily available to would-be terrorists, these will not be discussed in greater detail.

Bombs and related explosive devices have been employed in many terrorist attacks, especially in suicide bombings. Two major classifications are explosive bombs and incendiary bombs (e.g. Molotov cocktails). Terrorists also make use of letter and parcel bombs.

Few military bombs (other than those dropped by aircraft) are currently manufactured on the scale and diversity encountered in the Second World War. The exception to this generalization is the mine – both the anti-personnel and anti-tank mine. Mines can be adapted without too much difficulty with average combat-engineer experience. Some 300 different types of mines are embedded in the soil, killing tens of thousands every year.

Most bombs assembled by terrorists are improvised. The raw explosive materials are often stolen or misappropriated from military or commercial blasting supplies, or they may be made from fertilizer and other readily available household ingredients. Such assembled bombs are known as improvised explosive devices (IEDs).

Airline Baggage Security

Since September 11, 2001, airline security has become a major security focus. Travelers and their bags are now subjected to intense scrutiny. In a commercial airport, if a bag is not cleared by a human operator at the first and second levels of screening, it is diverted to level three, where it is subjected to additional screening with trace detection technology or computed tomography (CT), for example. In the United States, CT is the only technology certified by the Federal Aviation Administration

(FAA). However, it is slow and is therefore used only as a third-stage tool rather than as a technology for use in a high-volume, level-one system.

Trucks and Vehicular Threats

Several US states are considering the implementation of global positioning systems (GPS) or other technologies that will make it easier for law enforcement personnel to track and possibly divert or stop trucks that deviate from their planned routes – especially if they are carrying hazardous or explosive materials, or if they are operating in the vicinity of a nuclear power plant or other possible target such as a crowded building or a public event. The trucking industry is supportive of such measures and US groups such as the American Trucking Association have sought to develop new programs for screening and performing background checks on cargo truck drivers.

Nuclear Weapons

Radiation is a form of energy that is emitted from man-made devices such as microwave ovens, x-ray machines or nuclear bombs. Relatively low levels of background radiation also come from the sun and outer space as well as from uranium and other radioactive elements that are present in varying amounts in the earth's soil. These sources contribute to trace radioactivity that may be breathed or consumed when drinking water or eating food. Radiation that enters the body through the mouth, skin, lungs or mucous membranes results in internal exposure whereas external exposure involves radiation that does not penetrate beyond the skin. Radiation exposure is typically measured in units called "rem" or "sievert." (One sievert is equal to 100 rem.) According to the Centers for Disease Control based in Atlanta, Georgia, an average individual in the US is exposed to approximately 0.33 rem annually. Eighty percent of a typical person's exposure in the US comes from natural sources and the remaining 20 percent comes from medical x-ray devices and other man-made sources.

Albert Einstein's theory of relativity forms an important part of our fundamental understanding of atomic physics. Enrico Fermi, Otto

Hahn, Fritz Strassmann and Lise Meitner Key all made discoveries concerning atomic fission in the 1930s. The work of these pioneers laid the groundwork for the development of nuclear weapons development in the 1940s. The first known atomic bombs were built and tested under the Manhattan Project in the United States, and the weapons constructed as a result were deployed in the Japanese cities of Hiroshima and Nagasaki as a means to end the Second World War.

The Cold War began shortly thereafter when the USSR detonated an atomic bomb in 1949. In the 1950s, many concerned people and organizations began building bomb shelters with thick concrete walls and/or lead shielding. The first intercontinental ballistic missiles (ICBMs) were deployed by the United States in 1958 during the Korean War.

In the 1960s, France and China joined the so-called "Nuclear Club" and the Cuban Missile Crisis nearly instigated a nuclear war. In 1968, the Vietnam War began and the concerns and protests of many US citizens gave rise to the Non-Proliferation Treaty (NPT), in which the nuclear powers were to pledge complete nuclear disarmament. The signing of the SALT I and Anti-Ballistic Missile (ABM) treaties in the early 1970s, was followed by the SALT II agreement drawn up at the end of the Vietnam War. Then, the tragic Three Mile Island accident occurred, leading to a renewed awareness of the dangers of nuclear energy production.

At the height of the Cold War, President Reagan's administration engaged in an effort to amass large amounts of nuclear arms. In 1985, Israel announced that it possessed as many as 200 nuclear weapons. By the end of the 1980s, *glasnost* resulted in a mostly peaceful revolution across the former-Soviet Bloc and the Cold War was brought to an end. Nuclear disarmament in the former Soviet Union was boosted when Ukraine, Kazakhstan and Belarus chose to give up their nuclear arsenals. However, other states did not follow suit and by the end of the decade, India and Pakistan had begun testing nuclear weapons.

Radiation can affect humans in various ways. The type and amount of radiation that reaches a person's body, the route of exposure (whether internal or external), and the duration of the exposure determine whether or not adverse health consequences will result. Often, such adverse effects may not be observed for many years. While low doses of radiation may cause mild symptoms such as reddening of the skin shortly after exposure, this may also increase the risk of developing cancer several years later. Exposure to very large doses of radiation following a nuclear bomb detonation or nuclear power plant accident may cause death within a few days or months.

Potential nuclear terrorist attacks could involve the use of small radioactive sources with limited radiation potential, or large-scale nuclear detonations that affect very large areas. As with radiation from any other source, the consequential effects of radiation from a nuclear terrorist attack will vary according to the type of radiation and the distance between the source and any humans in the vicinity of the attack. If a nuclear bomb is detonated, shock waves from the blast itself or debris thrown from the blast could cause instant death in large numbers. Anyone who looks directly at the blast could experience temporary blindness or even severe and permanent retinal damage. In addition, many people are likely to be affected by short- and long-term health effects following the radiation.

There are three ways of protecting or reducing one's exposure to radiation:

- *Decrease* the amount of *time* spent near the source of radiation.
- *Increase* the *distance* between the radiation source and any humans.
- *Increase* the *shielding* between you and the radiation source.

Any substance that creates a barrier between humans and the radiation source may act as a shield. Certain types of radiation can permeate all but the most significant shielding (e.g. concrete walls that are several feet thick), whereas other types of radiation are shielded by something as thin as a plate of window glass or even a person's own skin. Being inside a building or a vehicle can provide shielding from some kinds of radiation. Preventive measures such as the establishment of adequate shielding could potentially reduce or eliminate exposure to radiation from a nuclear terrorist attack.

Many are concerned that the worldwide response to the threat of nuclear terrorism since September 11, 2001 has been severely inadequate. Although most military and national security leaders acknowledge the threat of nuclear weapons use by terrorists or even radical governmental entities, relatively few active steps have been taken to thwart would-be nuclear terrorists from obtaining weapons-grade nuclear materials and bomb-making hardware. While the danger of nuclear bomb construction, transport and deployment may be more significant now than it has ever been since nuclear weapons were first developed during the Second World War, little has been done to implement new systems for sensing or detecting nuclear weapons. Some experts believe that significant efforts and funding should be focused upon the development of nuclear detection systems.

The following seven steps were proposed in a 2002 report titled *Securing Nuclear Weapons and Materials: Seven Steps for Immediate Action*, published by experts at Harvard University and the Nuclear Threat Initiative (NTI) in Washington, DC. The report warns that these steps must be followed by the United States and Russia to prevent terrorists from obtaining nuclear weapons or their essential ingredients. The report recommends the following measures:

1. *Forging a Global Coalition to Secure Weapons of Mass Destruction*: Presidents Bush and Putin should seek to forge a global coalition to secure stockpiles of weapons of mass destruction (WMD) and their essential ingredients everywhere. Participants would pledge to secure and account for their own stockpiles to stringent standards, cooperate to interdict WMD theft and smuggling, share critical intelligence on these threats, and prepare to respond to WMD threats and attacks.
2. *Appointing One US and One Russian Official to Lead the Respective Countries' Efforts to Secure Nuclear Weapons and Materials*: Today, there is no senior official anywhere in the US government with full-time responsibility for leading and coordinating the range of efforts related to securing nuclear weapons and materials. President Bush and President Putin should each appoint senior officials, reporting directly to them, with no other mission.
3. *Accelerating and Strengthening Security Upgrades for Warheads and Materials in Russia*: The United States and Russia should jointly set a target of accomplishing all "rapid upgrades" of security and accounting for warheads and materials within two years and comprehensive upgrades within four years, and take a series of steps to build an accelerated partnership to achieve that goal.
4. *Launching a "Global Cleanout and Secure" Effort to Eliminate or Secure Stockpiles of Weapons-Usable Nuclear Material Worldwide*: A new program should be established to provide targeted incentives to

facilities worldwide to give up their weapons-usable nuclear materials, and to carry out rapid security and accounting upgrades wherever insecure nuclear materials remain.

5. *Leading Toward Stringent Global Nuclear Security Standards*: The United States should join with Russia and other like-minded states with substantial nuclear activities in making a politically binding commitment to meet a stringent, agreed standard for security and accounting for all their nuclear material and facilities, military and civilian – and to encourage others to do the same.
6. *Accelerating the Blend-Down of Highly-Enriched Uranium*: The Bush Administration should begin negotiating with Russia an accelerated approach to destroying Russia's excess bomb uranium, in which tens of tons of additional material would be blended and stored each year, for later sale. Congress should appropriate approximately US$50 million to fund the first year's accelerated blending.
7. *Creating New Revenue Streams for Nuclear Security*: New revenue streams should be developed that can supplement ongoing government expenditures for securing nuclear weapons and materials in the former Soviet Union, such as a "debt for nonproliferation" swap, or a set-aside of revenues from spent fuel imports, if an acceptable approach to such imports moves forward.

John P. Holdren, who is the Theresa and John Heinz Professor of Environmental Policy at Harvard's Kennedy School of Government, and one of the report's co-authors, observed, ahead of the Bush–Putin summit held in May, 2002:

> At their upcoming summit, Presidents Bush and Putin should complement their nuclear arms reduction agreement with an accord to sharply accelerate our countries' efforts to secure and account for all their stocks of nuclear weapons and materials, and invite other countries to join them in a global coalition to do the same.

Another co-author, Matthew Bunn, commented as follows:

> We have the technology to secure and account for all the world's nuclear weapons and potential bomb material, keeping them out of terrorist hands. We need sustained political leadership and resources to get the job done.

Bunn is a senior researcher in the Kennedy School's Project on Managing the Atom (MTA), which produced the report with the support of the Nuclear Threat Initiative (NTI).

The report emphasizes that nuclear weapons and materials are located at hundreds of military and civilian facilities in dozens of countries, with security conditions varying from excellent to appalling, and with no binding global security standards in place. At some facilities, security is provided by no more than a single night-watchman and a chain-link fence. As a result, documented thefts of weapons-usable nuclear material continue to occur – such as the seizure of nearly a kilogram of highly enriched uranium in the former Soviet state of Georgia in April 2000. While the United States and Russia are working together to secure and account for Russia's deadly Cold War nuclear legacies, to date even initial "rapid upgrades" – such as bricking over windows or piling heavy blocks on top of material – have been accomplished for only 40 percent of the potential bomb material in Russia, and less than one-seventh of Russia's stockpile of highly enriched uranium has been destroyed. The report calls for a drastic acceleration of these efforts, and outlines the means by which that could be accomplished.

Former Senator Sam Nunn, co-chairman of NTI warns:

> Terrorists are racing to get weapons of mass destruction. We should be racing to stop them. Believing that the recommendations in this report would help get us moving at a pace to win that race, NTI commends this report to policymakers in the United States and Russia. The time to act is now. President Bush and President Putin seem to understand this, but their challenge at this summit is to get their own teams heading in this direction.

Preparing for Disasters

In 'Dark Winter' – a simulated terrorist attack involving smallpox – the dramatic event developed into a worldwide epidemic within a period of weeks. The exercise clearly demonstrated the devastating effects of a bio-terrorist attack that used a contagious and lethal biological weapon of mass destruction. Local response and containment proved unsuccessful and the medical-consequence management efforts were overwhelmed.

Even with a national and implied international effort to control the disaster, a pandemic ensued.

Emergency preparedness is an essential requirement of national security. In order to protect communities both at the national and international level, efforts must be made not only to prevent such crises but also to respond effectively whenever they occur.

Crisis response to weapons of mass destruction will immediately involve local authorities and resources but will quickly require national and possibly international coordination and support.

The consequence to be managed in bio-terrorism and with most weapons of mass destruction is largely an issue of mass casualty care. Most healthcare systems would be woefully inadequate to manage a significant terrorist attack resulting in mass casualties. A coordinated surveillance, identification, containment, communication and response system would be necessary to minimize the effects of such an attack. The essential facets of such a system would include:

- Communications unhindered between local first responders and authorities, regionally, nationally and if necessary internationally.
- Integrated communications among detection units, laboratories, first responders and healthcare facilities.
- Improved detection equipment and enhanced laboratory detection; identification and diagnosis capabilities.
- Coordinated local and national medical surveillance for "real time" symptom and disease analysis.
- Accelerated specialized training of first responders and essential personnel.
- Improved medical and physical barrier-protection equipment for first responders and facilities.
- Ensured, timely access to medication and vaccines for the treatment of mass casualties.
- Decontamination facilities at all hospitals and treatment centers.
- Enhanced "surge" bed capacity and alternative mobile medical units and sites.
- Improved information technology for post-attack record-keeping including bed status and patient-tracking systems.

Conclusion

Though terrorism is apparently as old as the human race, it is now that we are challenged to work together as one world community to confront this ancient evil with new resolve and methodology. Nations have battled terrorism in their own lands, in their own ways. Now is the time to condemn and confront terrorism with one international voice and combined dedication. Unfortunately, there is always the likelihood of disputes, political conflicts, military actions and war. However, is it also true that these must escalate to terrorizing the innocent with criminal deadly intent? Recent actions and resolve among many nations of the world seek to intervene to prevent weapons of mass destruction from continuing to terrorize the citizens of our planet.

Nations are beginning to work together differently to combat terrorism. For instance, Germany and Italy have both encouraged the European Union to step up the fight against terrorism. In the fall of 2002, the United States, Germany, Czechoslovakia and Kuwait participated in military exercises focused on the consequences of an attack with VX nerve gas. These simulations can be highly effective in training across borders where weapons of mass destruction easily pass. Interoperability of communication and medical systems and equipment would dramatically enhance international competence in the management of mass casualties and the consequences of weapons of mass destruction.

The international community has a long history of pacts and agreements dealing with counter-proliferation. The Geneva Protocol of 1925 prohibited biological warfare. In 1975, the Biological Weapons Convention (BWC) banned the production, acquisition and stockpiling of biological weapons. Unfortunately the treaties and conventions lacked good mechanisms to ensure enforcement. In 1995, there was an indefinite extension of the Nuclear Non-Proliferation Treaty. In 1996, the Comprehensive Test Ban Treaty (CTBT) was signed. In 1997, the Chemical Weapons Convention (CWC) entered into force with an intrusive verification program. The Biological Weapons Convention may be at this time the least enforceable, although biological weapons certainly pose an equal and arguably the most serious threat to world peace and security.

The international community has demonstrated a growing commitment to the counter-proliferation of bio-terrorism and weapons of mass destruction, but this effort will require steadfast dedication if we are to win. It is no longer a problem just for India and Pakistan, Ireland and Britain, Israel and Palestine, the United States, the Western world, the Middle East or the Far East, or any single region or people. Although terrorism and our efforts to defeat it are at times fragmented, regionalized, politicized and personalized, terrorism is in fact truly a scourge affecting the entire world. It is a universal threat that will require unprecedented international cooperation if we are to find a global solution.

Section 4

Biotechnology and Industry

7

Investing in the Biotechnology Industry: The Role of Research and Development

Andrew Greene

Much of the discussion so far has been focused on biotechnology issues that are the subject of public and media debate – the human genome, genetically modified crops, cloning and bio-terrorism. For variation, this chapter concentrates wholly on the business and commercial aspects of making human medicines. In other words, making biotechnology products that interest both investors and buyers.

At the outset, the following are some basic points that need to be clearly stated:

- biotechnology is not only a science but also a business;
- demand for biotechnology products is expected to keep increasing;
- due to market specialization, small companies will do most of the drug research;
- drug research is a risky and expensive process;
- venture equity capitalists are likely to be the financiers of biotech companies;
- careful and canny venture investors can make premium returns as in the past.

Biotechnology is a broad term and its precise meaning depends on the person seeking to define or explain it. In the context of this chapter, the term is used to refer to the business of developing new human medicines. More specifically, this chapter focuses on inventing and commercializing new, proprietary drugs.

Drugs fall into two categories; small molecules and large molecules. Traditional drugs are small chemical molecules, made of carbon and other elements, and they work well because they are synthetic and alien to the human body. The body takes longer to break these chemicals down, thus allowing them to go straight in and do their job quickly. Among the best known of the small molecule drugs are aspirin, Valium and Prozac.

In contrast, large molecules are products of nature. They are large because they tend to be proteins (which are big, carbon-based molecules) comprising thousands of atoms, made by the human body. Insulin, Factor VIII and DNA itself are examples of large biotechnology "drugs." Doctors often need to inject them because they are harder to get into the body, which recognizes parts of them immediately and starts to break them down as soon as they pass the lips. Regardless of whether they are large or small molecules, all these products have dramatically improved human welfare.

Despite the medical advances in the last century, there are still huge unmet needs for cures for many of the major diseases (see Figure 7.1). Many medicines only treat symptoms. Statistics provided by the American Cancer Society indicate that in the United States, one of the richest and best-treated populations on earth, over 1,500 people die of cancer every day, while one person dies of heart disease every 33 seconds.

According to the PhRMA 2000 Industry Profile, the market is significant, with sales of treatments for the 12 major diseases, plus productivity costs, amounting to US$645 billion in 2000. Some diseases, such as tuberculosis and polio, are making a comeback, and with 30,000 disease mechanisms in the human body, scientists are continually discovering new gene targets from which they can design treatments. Even though, at the current time, more medicines and technologies are available than at any time in the past, there still is much to be done, and many opportunities to be seized, in order to improve medical welfare.

Figure 7.1
Growth Forever: Unmet Needs

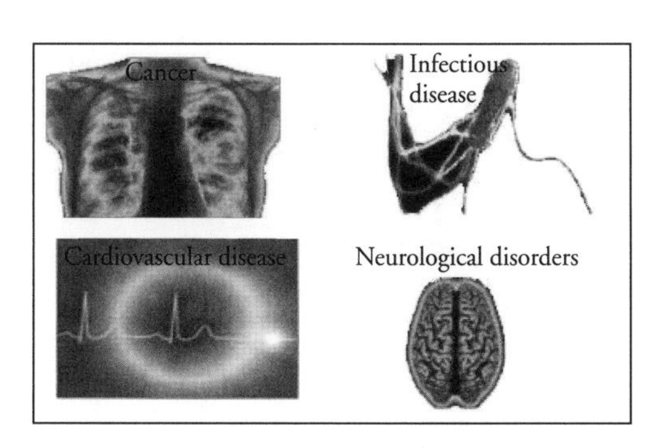

- New gene targets
- 30,000 disease mechanisms
- Medical sales per head increase 20% faster than GDP per head
- Unstoppable demand
- Governments control, markets create

Source: Merlin Biosciences Analysis.

Analysis tells us that the market for human medicines will grow at an accelerating rate as long as human wealth continues to increase. The practical implication of this statement is that healthcare demand will continue to increase indefinitely in the future. In the past, human civilizations have made themselves steadily wealthier for as long as history has been recording their achievements. We stand at the peak of human wealth and health, and even the poorest demographic quartile today lives better and longer than some kings did a few hundred years ago. In the long run, human progress is continuous and without limit. A few hundred years ago, our ancestors could not even imagine the levels of our power, productivity and mobility today.

Figure 7.2
Drug Research and Development

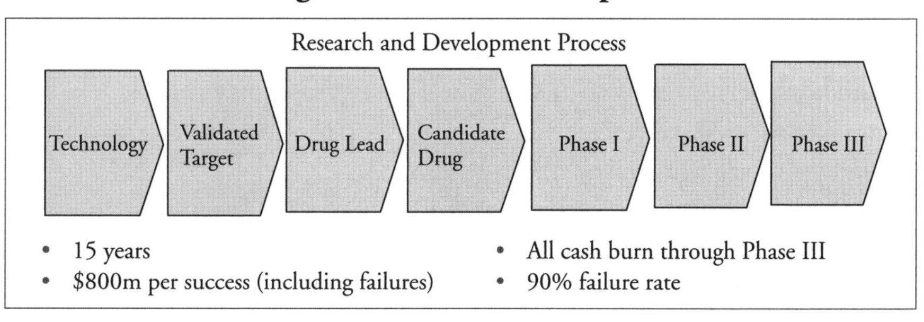

Source: Merlin Biosciences Analysis.

Hard data indicates that as the affluence of a population increases, so does its demand for medicine. In fact, the demand for medicine per head increases faster than the increase per head in general wealth. In a poor country, the average person might spend just a few dollars per year on medical care and these few dollars might be less than 1 percent of his income. On the other hand, in a rich country, the average person might spend several thousand dollars on medical care, and this could represent 8 or 10 percent of his income. Drug sales will grow to meet this unstoppable demand and people will spend a greater proportion of their disposable income on their health. Medical sales per head will outstrip the growth in income per head for a long time to come. There are and will continue to be variations in different countries, depending on the restrictions on pricing and competition that governments adopt, and whether they collect medical taxes. These constraints will either understate or overstate the true demand for medicine, but cannot change the underlying reasons why demand is unstoppable.

In Europe, the national government of each country controls the prices and supplies of drugs. In addition, tax money is taken to buy drugs for their taxpayers and dependents. In the current recession, most governments are trying to restrict their healthcare budgets. Analysis indicates that this move will also boost medical demand because people will work around the state system to get the desired healthcare. By circumventing the state system, people will tend to spend more and find that their money goes further. As long as governments do not expand their bureaucracies further into healthcare, the demand for drugs will grow.

Just as in other industries, companies in the medical industry are specializing more and more. Investors demand that companies concentrate on those aspects of the business that are capable of generating the greatest shareholder returns. For big pharmaceutical companies, their marketing infrastructure is becoming their greatest and best-protected asset, because it comprises thousands of specialist salespeople with profitable access to the doctors who prescribe medicines. These companies are so powerful at marketing that they can sell a third-class product far better than a smaller company can sell the first-class product.

At present, companies in the medical industry are moving further and further away from the fields of drug manufacturing, and drug research and development. They buy the inventions of other companies, pay third parties to make the drug, and then sell it themselves. In the same way as automobile manufacturers, they are effectively buying parts and component systems, but eventually selling the completed product. These marketing giants need products to sell. Many of these products are still invented in-house, but as more drugs lose their patent protection, it is imperative for these companies to find additional sources of new drug inventions, and the main outside supplier is the biotechnology industry.

Figure 7.3
Generic Exposure

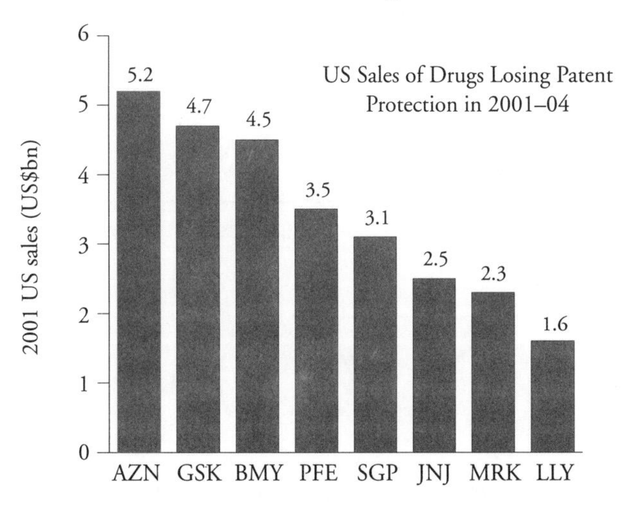

Note: AZN: Astra Zeneca PLC; GSK: GlaxoSmithKline PLC; BMY: Bristol-Myers Squibb; PFE: Pfizer Inc.; SGP: Schering-Plough; JNJ: Johnson & Johnson; MRK: Merck & Co; LLY: Eli Lilly & Co.

Source: S.G. Cowen.

Drug discovery is not only a long process but is also both expensive and risky. It can take a pharmaceutical company US$500 million and more than a decade to get one new medicine from the laboratory to the patient. Only five in five thousand compounds that enter pre-clinical testing make it to the stage of human testing, and only one of those five is finally approved for sale. In their bid to maintain the expected levels of double-digit sales and profit growth, the major drug companies have become increasingly dependent on biotechnology companies for platform technologies and to fill their product pipelines. Sixty-five percent of US New Drug Applications are for compounds that originated in biotechnology companies.

Figure 7.4
The Importance of Biotechnology

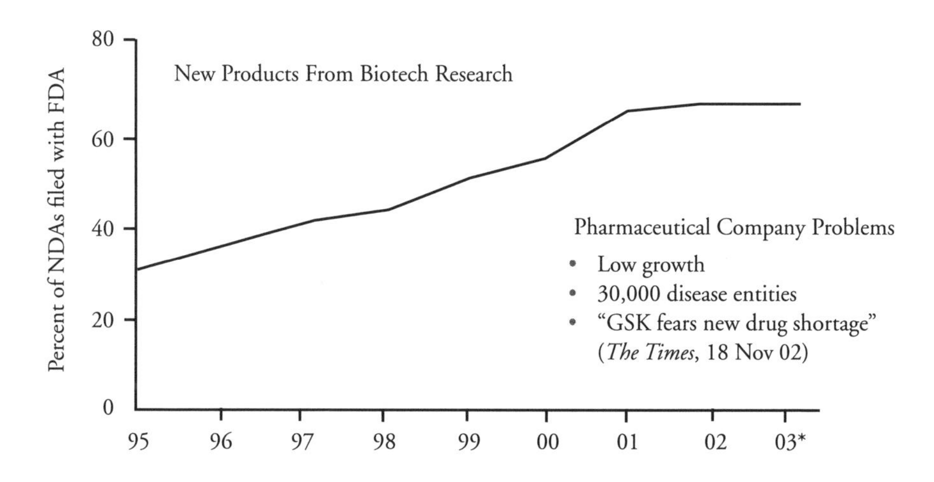

Note: *Projected.
Source: *Ernst & Young Global Biotechnology Report 2002.*

To fill their product pipeline, the major companies have gone shopping. Johnson & Johnson (JNJ) is a good example. As the fastest growing among the top ten pharmaceutical companies worldwide, JNJ's growth is coming from acquisition. Three of its four top drugs came from the research laboratories of biotechnology companies and 40 percent of JNJ's sales now come from in-licensed products. Five of the most advanced projects in the company's pipeline have come through acquisition or collaboration.

Pharmaceutical companies recognize that the products found in the later stages of the pipelines of several biotechnology companies have a real value and they are increasingly prepared to pay for these products. In 2002, for the first time, the total amount paid upfront by major pharmaceutical companies for late-stage products outstripped the total amount paid for all early-stage deals combined.

Figure 7.5
Rising Prices for Late-Stage Deals

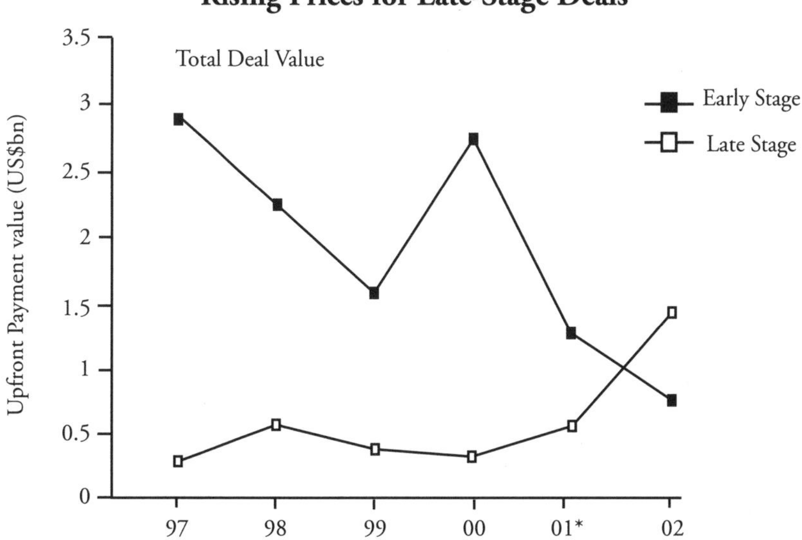

Note: Late stage includes deals signed in Phase II, III and after filing for approval and only those with disclosed upfront payments; 2002 data is as of September 15, 2002 annualized.
Early stage excludes preclinical deals; includes only deal with announced value; 2002 data is eight months annualized.

*Excludes BMY–Imclone deal worth US$1.2bn.
Source: *Windhover's Strategic Transactions Database.*

Equity risk capital is now a major source of finance for the high-risk research and development stage of human medicine. Biotechnology venture capital is big business. In 2001, general partners made US$3.8 billion worth of venture investment in the US and Europe into three thousand private biotechnology companies. Venture capitalists are interested in investing because they can purchase the rights to private intellectual property that will generate sales or royalties. The high risk and uncertain nature of their investment is counterbalanced by levels of return that are double that of public equity returns. At exit, they typically achieve three to five times return on their original capital. The appetite of major pharmaceutical companies for the attractive product pipelines of biotechnology companies keeps the trade sale exit route open even when the IPO window might otherwise be closed. This is another contributing factor for venture capitalists' interest in this sector.

Figure 7.6
Venture Capital Equals Research and Development

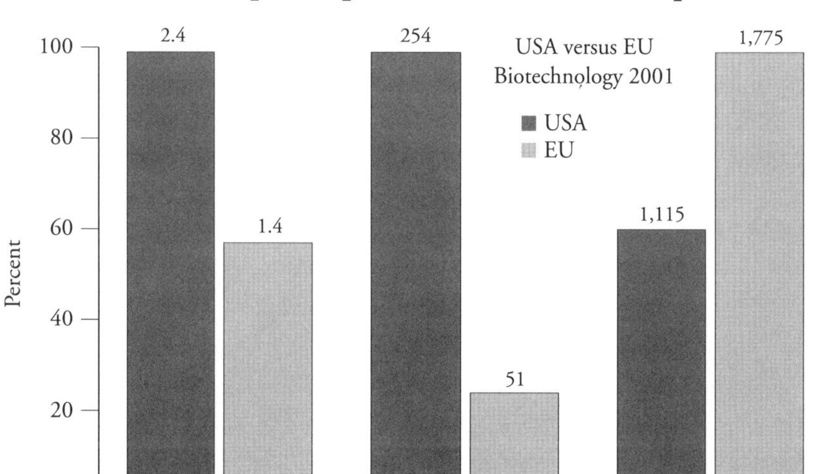

- Private intellectual property, *not* "common property"
- 3x to 5x investment returns
- Favorable capital gains tax (UK)
- Trade and IPO exits

Source: E&Y 2002; Merlin Biosciences Analysis.

These are tough times for attracting investment. After experiencing the greatest financial bubble ever, even bigger than the South China Seas or the Tulip bubble, confidence is at an all-time low. Cash is king and investors are loath to let go of their money. The good news is that in biotechnology investment, this low level is part of an historical cyclical trend. In the glory days, the public stock markets were typically open for twelve to fifteen months before they went through a downturn, usually for two to three years. Since 1987, the industry has gone through three of these cycles. Although today's downturn is worse, the industry has a stronger foundation to fall back on. The volume of capital committed at the peak of the last cycle created expert investors. There are now over forty dedicated biotech funds and it is unlikely that such expertise will abandon the sector.

Figure 7.7
Crisis of Confidence

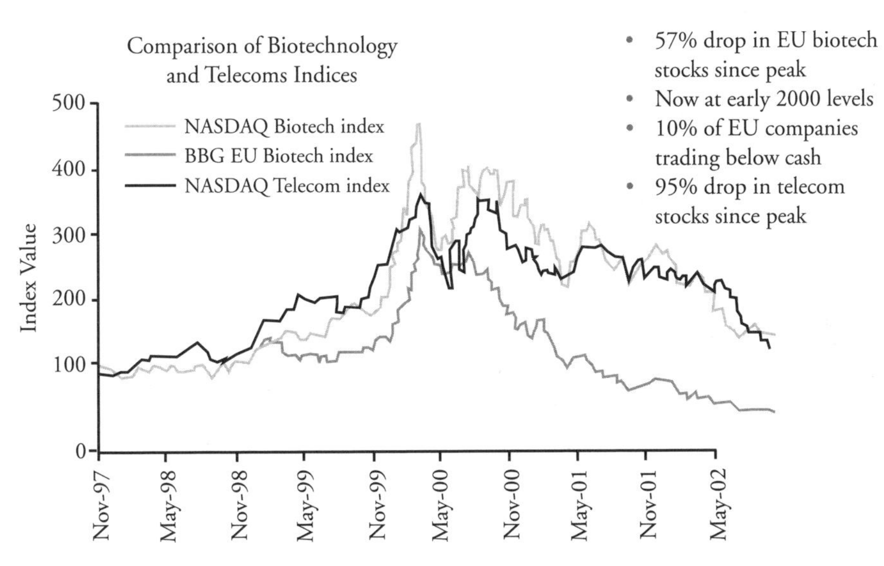

- 57% drop in EU biotech stocks since peak
- Now at early 2000 levels
- 10% of EU companies trading below cash
- 95% drop in telecom stocks since peak

Source: Bloomberg.

Figure 7.8
Biotech Investment Cycle

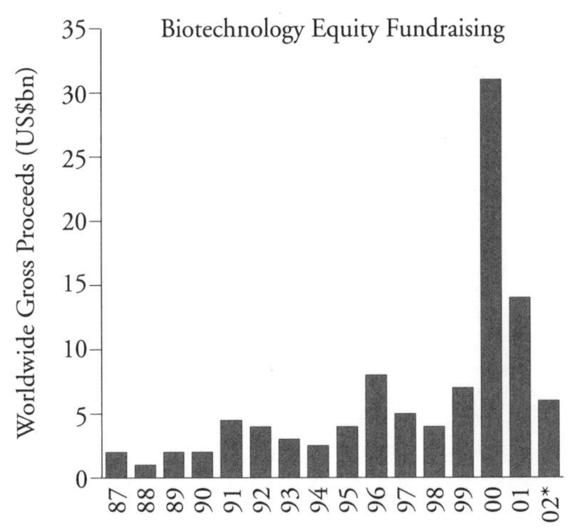

- Glory days
  - Markets open 12–15 months
- Gory days
  - Markets closed 30–36 months
  - Scandals, failures, pessimism
- Today
  - Private tech values greater than public
  - Pre-IPO values decreasing
  - 40% down rounds (versus 72% for IT)
- But, a stronger foundation

Note: Excludes corporate alliances
*2002 October YTD annualized = US$5.2bn
Source: *BioCentury*; Brobeck Hale & Dorr, November 2002.

Consumers have always demanded and will continue to demand that they live well. Drug sales in 1999 of US$20 billion will swell to over US$70 billion by 2009, and the sector might perhaps be valued at US$1 trillion by the end of this decade. Even today, there are 4,000 biotechnology companies employing over 223,000 scientists and staff.

Figure 7.9
Global Growth Opportunity

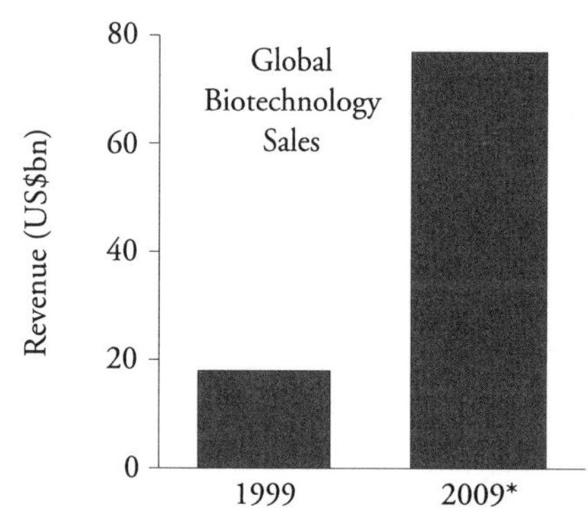

- Correction has happened:
  - US$600 billion valuation end 2001
  - US$300 billion valuation today
- Consumer demand: to live, to live well
- US$1 trillion valuation by 2010
- 4,000 companies
- 223,000 employees

Note: *Projected.
Source: *Ernst & Young 2002*; European Commission – *Life Sciences and Biotechnology: A Strategy for Europe 2002.*

Figure 7.10
Cash-Starved Biotechs

Announced Headcount Reductions	*Percent*	*Date*
Paradigm Genetics	20	Apr 02
Hyseq	39	May 02
Large Scale Biology	31	Jun 02
Bio Transplant Inc.	23	Jul 02
Alkermes	23	Aug 02
Entre Med	25	Aug 02
Targeted Genetics	25	Aug 02
Dyax	16	Sep 02
Genaissance	20	Sep 02
Immune Response	67	Sep 02
Xanthon	100	Sep 02
Ax Cell Biosciences	75	Sep 02
DeCode	30	Sep 02
Neo Therapeutics	52	Various

Source: *IN VIVO*, October 2002.

However, as usual, there is a catch. Today, every biotechnology company must have both a revolutionary technology and a revolutionary commercial application for it. The days when a scientist with a bright idea could have a US$10 million valuation at his seed round are over. The post-boom biotechnology business model demands a steep development curve and a short time to revenue and profits. If that does not happen, early backers will see their investment wiped out by price-sensitive investors at a later-stage financing, which must be completed under pressure because the prospects of the company are weaker.

Figure 7.11
Investment Reality

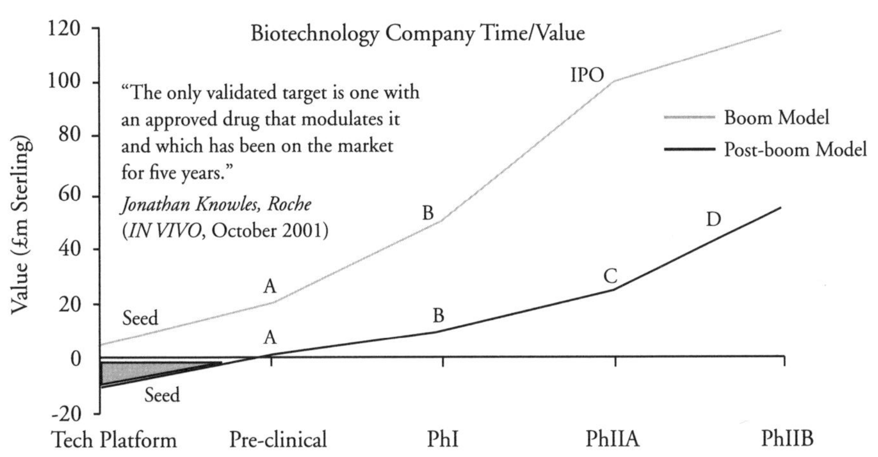

Source: Merlin Biosciences Analysis.

Thus, the companies that invest in research and development must have products that substantiate a realistic valuation. Jonathan Knowles of Roche was quoted by *In Vivo* of October 2001 as saying, ". . . the only validated target (which researchers use to design drugs and which exchanged hands for large amounts during the boom) is one with an approved drug that modulates it and which has been on the market for five years." Few research and development companies will have such an ideal opportunity, but they must if they want financing for their work. The bar is much higher now.

Figure 7.12
Private Finance Outstrips Public Finance

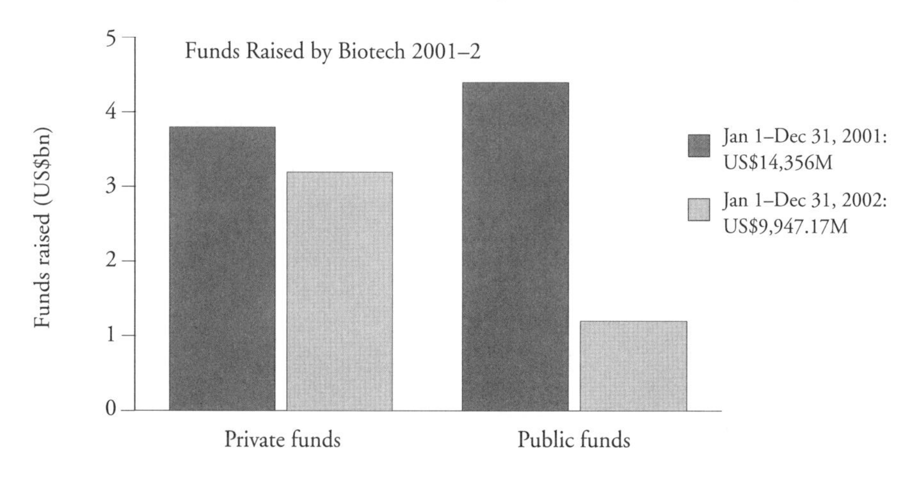

Source: Bio World.

Figure 7.13
The Future of Biotech Research and Development

- Demand for new drugs will increase indefinitely
- Profit is the prime motor of R&D productivity
- The largest source of R&D capital, the pharmaceutical industry, is cutting back
- Expert venture investors will lead R&D growth
- The highest risk and highest returns will remain with private equity

Source: Merlin Biosciences Analysis.

In conclusion, it is worth restating the points already outlined at the beginning of this chapter. Biotechnology has a great future, and its business model is sound. Humanity's demand for new drugs will increase indefinitely. Private property and profit, as long as they remain legal, will continue to be the motors that drive drug research and development productivity. The largest source of research and development capital, the pharmaceutical industry, is cutting back. Therefore, expert venture investors will need to lead research and development growth, which they will do willingly despite the high risks involved, because of the high returns anticipated in this sector.

8

Biotechnology and our Material Future

*John Pierce**

Ever since the discovery of the structure of DNA in the 1950s and the elaboration of the manner in which that structure encodes the building blocks of living organisms, massive investments of time, intellect and capital have provided us with an improved understanding of biological principles and increasingly sophisticated tools for utilizing biological concepts in valuable ways. About thirty years ago, we began developing the capability to utilize molecular biology to specify – at a molecular level – the production of useful proteins and other products. The biotechnological revolution that began with these early experiments is now fully upon us, and has already had major effects on our lives by providing us with new medicines, foods and crops, and new ways to ensure the safety of our food supply and diagnose human disease.

The use of biotechnology for the production of chemicals and materials of industrial value is also emerging as a major economic force. Driven by continuing improvements in the efficiency of agricultural production (which provides the necessary raw materials), the exploding knowledge base of DNA structure, and more powerful biotechnological tools, the biological production of very large volumes of important chemicals and

* The author would like to acknowledge the substantive input of Drs Armando Byrne, Scott Cunningham and Vasantha Nagarajan in the creation and preparation of this chapter, and the creative input of Mario Chen in the preparation of the images used in the conference presentation.

materials from renewable resources is becoming a reality. These materials are finding applications in energy, transportation, clothing and housing – in short, in all the major industries that utilize chemicals and materials. In addition, we are learning to employ biotechnological approaches to generate very small structures with high precision for use in everything from medicine to electronics to precision coatings. So, from the very large scale to the very small, biotechnology is poised to make important contributions to our material future.

Despite the impressive advances, we are just beginning to learn how to incorporate biological principles into truly multidisciplinary approaches with engineering, chemistry, physics and material sciences – disciplines with long, successful histories of innovation. As we do so, we will improve our capacity to utilize renewable resources and renewable concepts in the production of materials, and will take a major step toward ensuring a sustainable future.

Biotechnology's Evolving Role in Materials Production

Humans have long recognized the power and utility of biology or biotechnology in their daily lives. Early examples of biotechnology can be found around the world; particularly, and perhaps not surprisingly, in the area of fermentation in dealing with the very real problem of food and beverage spoilage. Still more complete capabilities of biotechnology have only recently become available with the development of recombinant DNA technology. Concomitant advances in process technology, chemistry and classical engineering, and their integration with biology have enabled biotechnology to be applied in areas well beyond food, into uses in agriculture, pharmaceuticals, and energy and materials production.

Our modern societies are completely dependent on materials – for nutrition, health, energy and manufactured articles. These huge, trillion-dollar sectors consume enormous quantities of output from agriculture, oil fields and mines. During the twentieth century, we learned how to produce an amazing array of materials by transformation of petroleum into new chemical and polymeric materials. Billions of pounds of these materials are produced each year, and find use in almost everything we do. As articles of clothing, electronic devices, and housing and transportation materials, they contribute enormously to our well-being, keeping us warm in the winter and cool in the summer, as well as maintaining and preserving our food.

These benefits have not come without costs, and the rise of biotechnology has come at a time when concerns over humankind's impact on the global environment continues to be raised. Growing societal pressure is being applied to both governments and industry to seek alternative, cleaner processes to meet the growing needs for food, shelter and health. At the same time, the finite nature of our most widely used and non-renewable resource, fossil fuels, has resulted in growing calls to develop new resources for energy and materials production. Together these forces have culminated in a global discussion around industrial sustainability and the role of biotechnology in that quest.

Environmental Sustainability

A major element in all living cells is carbon, which is comparatively rare in the Earth's crust. Carbon, often referred to as the building-block of life, is present in all cellular molecules. Through evolution, living things have become quite adept at the degradation and synthesis of these organic compounds. At the global level, this balance between synthesis and degradation is played out in the carbon cycle. Light energy from the sun is used by plants, algae and certain bacteria to drive synthesis of cellular constituents from carbon dioxide. At the other end, certain microorganisms are largely responsible for liberating the carbon sequestered in these molecules into the atmosphere as carbon dioxide. Within this cycle of carbon, nature has provided humankind with many of the tools needed to build societies, tools that come in the form of carbon-rich raw materials.

Examples of such materials range from fuels, food and fibers to a large portion of the commodity chemicals, pharmaceuticals and non-durable goods. Before the use of fossil fuels, such as petroleum, coal and gas, societies relied on carbon-rich raw materials derived from renewable sources from agriculture and forestry. In the 1800s and 1900s, people started to tap into natural subterranean stores of fossil resources, prompting the gradual switch from biological raw materials to fossil fuel resources. Today, products derived from petroleum rather than renewable biological resources dominate large portions of the energy and organic chemicals markets. To be sure, the development and use of fossil resources has resulted in tremendous advances in the overall standard of living of people across the globe, but it has also contributed to undesirable consequences such as environmental pollution and global warming.

The carbon cycle in nature operates on the assumption that all biosynthetic organic compounds are biodegradable. With the advent of petroleum-derived organic chemical compounds, this assumption has been greatly tested. Not surprisingly, these never-before-seen chemicals make their way into the environment, where they can sometimes accumulate, prompting questions of environmental concern. At the same time, continued use of fossil fuels for energy production has resulted in the release and accumulation of carbon dioxide in the atmosphere – carbon that, prior to its use, was hidden away in dormant reserves far beneath the ground. It is widely believed that this accumulation of carbon dioxide may be a major driver in the greenhouse effect, which may be resulting in gradual yet significant increases in global temperatures.

These environmental concerns have caused governments under societal pressure to legislate new local limits on what are considered acceptable levels of environmental impact. The global dimension to humankind's impact on the environment, however, has also caused countries to work together to find global solutions to these difficult and complex problems, prompting the first multi-nation meeting devoted to the environment in Rio de Janeiro in 1992. This, in turn, has caused industry in general to re-examine its processes with the intent of reducing the environmental footprint left behind from its practices. Within this broad environmental discussion, the term "sustainability" has emerged.

By the mid-1980s, mounting evidence of human impact on the world's environment and the sustainability of environmental quality led to a growing appreciation that resource and waste management needed to improve. These environmental concerns, coupled with the need for continued economic growth and wealth creation, led to the development of the Bruntland concept of sustainability, which defines sustainable development as "strategies and actions that have the objective of meeting the needs and aspirations of the present without compromising the ability to meet the needs of the future." Within an industrial context, sustainable development therefore looks towards "clean technologies" to lower resource consumption and to reduce or eliminate waste. It is here that biotechnology can be a powerful enabling technology for the production of clean industrial products and the development of environmentally benign processes.[1]

Modern-Day Biotechnology: From Cloning to Genomics

As indicated earlier, the early roots of biotechnology can be seen with brew masters and cheese makers who used the action of bacteria and yeasts to create new foods and beverages. More recently, modern-day products such as antibiotics, amino acids and enzymes have been produced via similar microbial-based fermentative processes. In these examples, biotechnology is a broad term used to describe various techniques for using the properties of living systems to make products or provide services. This term would therefore be applied not only to traditional activities such as making dairy products, bread or wine, but would also include selective breeding, or plant cloning by grafting, or the use of microbial products in fermenting. What distinguishes these examples of "traditional" biotechnology from what we will describe as "modern" biotechnology is not the principle of using various organisms but the techniques for doing so. In this second-generation biotechnology or what could also be referred to as molecular biotechnology, recombinant DNA technology takes center stage in enabling the precise design of biocatalysts – either whole organisms or specific enzymes – for industrial processes. These industrial processes not only involve the integration of molecular biology, genetics, microbiology, cell biology and biochemistry, but also incorporate chemical and process engineering in a way that develops the full potential of each of these processes.

Recombinant DNA technology, also referred to as gene cloning or molecular cloning, is an umbrella term that encompasses a number of techniques that enable the transfer of genetic information, in the form of DNA, from one organism to another. Although recombinant DNA technology developed from discoveries in a variety of different fields of study, the fundamental knowledge that made it all possible stems from an understanding of the structure and function of DNA. This started with the discovery in 1953 by Watson and Crick of the structure of DNA. This was later followed in the early 1960s with the deciphering of the genetic code, which led to an understanding of the individual units of genetic information called genes and to their relationship with the proteins they encode. A decade later, Cohen and Boyer utilized this information to perform the first cloning experiment, splicing together the DNA from two different organisms and creating the first recombinant DNA organism. The Cohen and Boyer strategy of gene cloning established recombinant DNA technology and led to profound technological developments that would significantly impact a number of biological disciplines, most of all biotechnology.

In the thirty years since those first experiments in gene cloning, recombinant DNA technology has exploded with technical advances, leading to the creation of powerful new technologies and disciplines. Among these are genomics, involving the molecular characterization of whole genomes; bioinformatics, the interdisciplinary field concerned with the assembly, storage, retrieval and analysis of computer-stored biological databases; and proteomics, which seeks to define the entire complement of cellular proteins. The combined force of these technologies and others have resulted in fantastic advances in our understanding of the world around us, the culmination of which is best illustrated with the recent completion of the Human Genome Project.

These same technological advances have also become powerful drivers in the application of biotechnology in industrial processes, where the development of suitable biocatalysts is key. Chemical processes are often conducted at high temperatures and high pressures, resulting in high energy utilization and the generation of unwanted by-products. Biotechnology offers new and cleaner synthesis routes by providing novel biocatalysts, which, by virtue of their origin, can often carry out their functions at ambient temperatures and pressures, with a high degree of specificity. These biocatalysts, living organisms or their catalytically active constituents, can be obtained from natural diverse habitats, or through modification of existing biocatalysts by genetic or physico-chemical methods. With microbes providing the greatest biocatalytic diversity and with only a fraction of the microbial population cultured and identified, several biotech companies, including Diversa from San Diego, California, have made it their business to "bio-prospect" the world's numerous environments for unique biocatalysts. At the same time, several companies, taking their cues from evolution, have developed new technologies, particularly in the areas of protein engineering and directed evolution, to generate great molecular diversity with the hopes of improving or creating new functionalities. When coupled with powerful screens, these technologies can provide tailor-made enzymes perfectly suited for the intended process conditions. At the cellular level, through metabolic engineering, new and/or improved biocatalysts can be designed by modifying specific biochemical reactions or introducing new ones to the cell.

Economic Drivers: New Products from Renewable Raw Materials

Whereas environmental concerns have led to the emergence of sustainability as a parameter on the industrial agenda, and advances in biotechnology have made the development of cleaner processes possible, it is the economic requirement that will ultimately determine the wider accessibility and adoption of biotechnology in the manufacture of materials, chemicals and fuels. The use of renewable resources is closely tied to the price of the fossil raw material they might replace. We live in a time where fossil-fuel-based feedstocks dominate, supported by the low cost of petroleum and by the well-established distribution network that permits the easy transportation of raw materials from the source to the consumer. And while the costs of some bio-based alternative feedstocks – for example, corn – are relatively low in price, the costs of processing these feedstocks for subsequent utilization make them uncompetitive versus the incumbent. Nevertheless, increasing prices for petroleum and questions regarding the longevity of these raw materials have resulted in increased interest and substantial research in the production of chemicals using renewable resources in the United States. As an example, the Biomass Research and Development Act passed by the US Congress directs the Department of Energy (DOE) to consider biomass as a source of raw sugars for chemicals and to lower the costs of bioethanol fuel. The DOE expects enzyme producers, such as Novozymes and Genencor International, to lead the way in finding new enzymes that will lead to improvements in costs for biomass conversion. More recently, companies, including Dow Chemical, Eastman Chemical, Rohm and Haas, and DuPont have obtained major federal grants to explore the use of alternate bio-feedstocks for the production of chemicals and fuel.

Potential sources for biological raw materials include agricultural and forest crops, and biological wastes. Agriculture generates some thirty trillion pounds of grain equivalents per year and even larger amounts of currently underused by-products. These are staggering amounts of material; more or less equivalent to the amount of petroleum we utilize each year. Over most of the past century, oil has been much cheaper than agricultural products. However, increased oil prices and improved agricultural productivity have tended to decrease the relative cost advantages of oil, and now basic raw materials from agriculture are on a more equal footing. Raw materials such as starches, celluloses and oils are already extracted from plants for the production of some biomaterials, chemicals and fuels. Fermentable sugars derived from corn, sugar beet and sugarcane are used in microbial fermentation processes to produce ethanol, acetic acid, amino acids, antibiotics and other chemicals. Yet the growth of the bio-based chemical industry will depend on further improvements in processing technologies to tap into cheaper sources of fermentable sugars such as the cellulose and hemicelluloses present in biological wastes.[2]

A step in the expansion of bio-based industrial production will require the scale-up of manufacturing capabilities. This may come in the form of "biorefineries" that will act as efficient processing plants that will produce a host of bio-based products. Similar to oil refineries, biorefineries are expected to push processing costs down over time, making their products more competitive on the market.[3]

I have described three principal drivers for industrial biotechnology processes; societal, technological and economic. Societal concerns over the environment and the future welfare of the planet, advances in biotechnology, and economic questions regarding raw material feedstocks, have all converged to make this an opportune time for the application of biotechnology in materials production. These developments have not gone unnoticed by governments and companies alike, where numerous partnerships have developed to make industrial sustainability a reality and to cash into this next technological revolution. In the next section, I provide some examples of how biotechnology has had an impact on industrial processes at different scales.

Biology as a Way of Making Things

While biotechnologically derived materials represent only a small fraction of the multi-trillion US dollar materials markets, the impact of biotechnology is clearly increasing and there has been tremendous progress in the last five years. A number of new processes are in the very late stages of pre-commercial activity, and the pipeline is filling up with additional examples. I will provide you with some examples of current commercial bioprocesses and new processes that are on the horizon to illustrate the enormous range of chemical and material transformations that are currently being pursued.

There are two primary ways in which biotechnology is used to produce new materials: via enzymatic conversion of a starting material or via microbial fermentation of a simple feedstock to a desired chemical through multiple conversions. (Although other approaches, such as the use of plants and animals to produce materials are also being investigated, the timeframes and economics of working with these organisms are such that enzymatic and microbial production are likely to be the major approaches for a long time.) Enzymes, the remarkably specific protein catalysts found in all living organisms, are increasingly being used as biocatalysts to effect difficult chemical transformations, and new technologies of enzyme evolution, protein engineering and selection allow scientists to generate improved biocatalysts that can work under a wide range of commercial operating conditions and on a wide range of commercially important starting materials. When very large quantities of materials are required, or when the task is to convert a readily available raw material such as glucose to a structurally dissimilar molecule – such as ethanol – microbial fermentation is used, and increasingly sophisticated tools for engineering microbes to produce different products are being brought to bear to expand the universe of industrial microbial transformations.

Industrial Enzymes

Enzymes found in nature have been used in a wide range of processes for a long time. In fact, the earliest references to the use of enzymes in cheese-making date back to Greek poems in 800 BC. In the early days, enzymes were used along with microbes in making cheese, sourdough, and in the manufacture of leather. Pure enzymes were commercially produced at the beginning of the twentieth century, and now most commercial enzymes are produced in engineered microbes. The estimated value of the industrial enzyme market in 2002 is US$1.5 billion. This is the market for the enzymes themselves, which are in turn used in a wide range of applications. Let me give a few examples of the use of enzymes for various processes that illustrate some of the benefits that can be derived from using these remarkable catalysts.

The major product of the industrial enzymes market is proteases for the detergent industry. Through the application of protein engineering to create altered enzymes, proteases that can operate at lower temperatures in household washing machines have been recently introduced. This provides a method for reducing the energy requirement without compromising on the cleaning functionality.[4]

Another example can be found in the textile industry, which has long struggled with the large requirements of water and energy and the resulting pollution associated with its textile wet processes. One of the most problematic steps in the process involves the "scouring" of cotton, which is performed at high temperatures and under strong alkali conditions. As a cleaner alternative, the biotechnology company, Novozymes, introduced the enzyme pectate lyase, which can carry out the same removal of cell-wall components from cellulose fibers, but at a much lower temperature and with less water utilization. For this, Novozymes received the US Presidential Green Chemistry Challenge Award.[5]

A third example is the use of nitrile hydratase in the synthesis of an intermediate for the DuPont herbicide, Milestone®. A starting material for the synthesis of this herbicide is 5-cyanovaleramide (5-CVAM). A chemical hydration of adiponitrile using manganese dioxide as a catalyst was originally developed to produce kilogram quantities of 5-CVAM, but significant amounts of the by-product adipamide (~20 percent) were produced at only 25 percent conversion of adiponitrile, and a difficult solvent extraction was required for the separation of 5-CVAM from unreacted starting material. Rapid deactivation of the manganese dioxide resulted in the production of 1.25 kg of catalyst waste-kg^{-1} 5-CVAM. In contrast, a "green" biocatalytic process was developed using the bacterial nitrile hydratase from *Pseudomonas chlororapis*. This bioprocess produced significantly less catalyst waste and by-products, and eliminated the product purification step due to the almost complete conversion of starting material to product. This process won an Industrial Innovation Award from the American Chemical Society (ACS).

Finally, one of the most successful examples of biocatalytic production of a commodity chemical is the conversion of acrylonitrile to acrylamide. Acrylamide is commonly used as the starting material for the production of a variety of chemical derivatives for use in the manufacture of polymers for paper treatment, flocculants, and enhanced oil recovery. The world's annual acrylamide requirement is over 180,000 metric tons. Mitsubishi Rayon Co., Ltd. (based in Tokyo, Japan) currently produces around 20,000 metric ton-year^{-1} of acrylamide using a third-generation biocatalyst, *Rhodococcus rhodochrous* J1, which was first isolated by Kobayashi and Yamada and developed for commercial use by Nitto Chemical Industries (now part of Mitsubishi Rayon Co., Ltd). Acrylamide is produced continuously from acrylonitrile at 10 degrees C in a series of fixed-bed reactors using polyacrylamide-immobilized J1 cells. Complete conversion of acrylonitrile produces acrylamide in ~99.9 percent yield at final concentration of ~50 wt percent in water, and the catalyst productivity is > 7,000 grams acrylamide-g^{-1} dry cell weight. In comparison to the traditional copper catalyst process, the bioprocess does not require heat or pressure, eliminates heavy metals, and produces less process wastewater than the chemical process. The acrylamide has a higher purity than that produced by the chemical process, which allows for the manufacture of higher-molecular-weight polyacrylamide. SNF Floerger (based in Saint-Etienne, France) has licensed the Mitsubishi Rayon process and plans to construct five new acrylamide plants, each with 20,000 tons of annual capacity, and Mitsui Chemicals also plans to begin biocatalytic production of acrylamide.[6]

As outlined above, enzymes have found their way into a number of different industrial products and processes. They can serve as components in end products, such as laundry detergents, or they can serve as biological catalysts in the industrial processing of textiles and specialty chemicals. In many cases, enzymes serve to replace chemical process steps, resulting in improved selectivities and product yields, and often leading to hard savings in resource consumption and reduced environmental impact. It is expected that with advances in protein engineering and directed evolution, the industrial uses of enzymes will continue to expand into other sectors of the world materials market. Still, as briefly discussed earlier, there exist commercial opportunities where very large quantities of a product are required, or where the conversion of a readily available and less expensive raw material to product requires a number of chemical transformation steps. In these cases, the use of enzymes as described above is no longer viable and microbial fermentation is needed. Several of these cases will be discussed next.

Fermentation-Based Bioprocesses

The earliest examples of whole cell microbial biocatalysis can be found in the production of foods or chemical additives in food applications. In addition to alcoholic beverages, this included food fermentations producing lactic acid in yogurt and sauerkraut, and acetic acid as in vinegar. As our knowledge of biology increased, so did our use of microorganisms to produce other materials, including fuels, commodity and specialty chemicals, and pharmaceuticals. One such case involves the production of riboflavin. Riboflavin or vitamin B2 was produced by a chemical method from glucose by Roche. The original method was first used in 1942 and involved six chemical steps from glucose. An intermediate method was installed in 1980 that involved conversion of glucose to ribose by fermentation and further chemical catalysis to yield riboflavin. In 2000, Roche installed a bioprocess that can produce 2,000 tons per year using a recombinant microorganism, *Bacillus subtilis*. The development of the complete bioprocess started in 1988 and involved metabolic engineering of *B. subtilis* so that glucose can be converted to riboflavin. A life-cycle analysis in this case shows that the bioprocess is more sustainable than the chemical process, mostly due to the use of renewable raw materials. Here is an example of a bioprocess replacing a chemical process after six decades – driven mostly by the environmental benefits.[7]

Recently, the problems concerning the global environment and solid waste management have created much interest in the development of biodegradable polymers. Several companies, such as Metabolix in Cambridge, Massachusetts, have sought to produce such materials using the biologically derived polymer, polyhydroxyalkanoate (PHA). PHAs are polymers of hydroxyalkanoate, which are accumulated in various microbes as energy and carbon storage material. There has been sporadic interest in these polymers since the first observation of polyhydroxybutyrate was made in 1926. Recently, much progress has been made through genetic engineering to produce large quantities of PHAs during fermentation from natural sugars. Metabolix has engineered a strain of *Escherichia coli* to produce PHA from commercial-grade corn sugar and recently reported achieving production of its PHA at the 50,000 liters scale with high production rates and titers.

Polylactides (PLA) form another interesting group of biopolymers that have recently attracted a fair amount of attention as a consequence of Cargill-Dow's efforts to produce the polymer from annually renewable resources. PLAs are aliphatic polyesters, which are composed of lactic acid monomers. They are thermoplastic and biodegradable. Cargill-Dow has been able to reduce the cost of manufacture of lactic acid through the development of a fermentation process to make the two chiral isomers of lactic acid from glucose. Under the brand name "Natureworks PLA," Cargill-Dow will be able to produce 140,000 metric tons of PLA at its recently opened plant in Blair, Nebraska when at full capacity. NatureWorks PLA is being targeted for use in the fiber and packaging markets around the world.[8]

Finally, another example of a high-volume commodity chemical is 1,3-propanediol, or 3G as it is known among chemists and polymer scientists. As a monomer, 3G has been known for a long time by synthetic polymer chemists. It can be used to make a polyester with very attractive properties. Although both Shell and DeGussa developed chemical routes for the synthesis of 3G, the chemical methods for synthesis of 3G were never sufficiently economical to provide a strong incentive to develop a polymer platform based on 3G. In the mid-1990s, polymer chemists at DuPont challenged their biological colleagues to develop a biosynthetic method. While biological production of 3G from glycerol has been known in the literature for more than five decades, past efforts to produce glycerol by fermentation were not commercially viable even though several attempts had been made. Using modern biotechnological approaches, DuPont, in collaboration with Genencor International, developed a biocatalyst that can convert glucose to 3G at high titers and productivity. Numerous modifications to the host organism had to be made, along with the introduction of foreign genes in order to produce 3G at the extremely high concentration necessary for commercial success. The success of this scientific endeavor has shown that it is possible to dramatically channel the carbon flow in microbes to make commodity chemicals by fermentation. Thus, 1,3-propanediol has the potential to be one of the high-volume commodity chemicals of the future produced by fermentation.

However, traditional fermentation already produces large volumes of ethanol by fermentation. Large-scale production of ethanol from biomass can provide cleaner fuels from a sustainable energy source. The current level of production of bioethanol in the US is 1.6 billion gallons and this is anticipated to reach 5.6 billion gallons by 2008. The European

Union regulations on biofuels aim to raise the market usage of bioethanol from 2 percent of fuel used in 2005 to 5.75 percent by 2008. There is potential for worldwide production of bioethanol to be around two billion tons per year in the next decade. Most of the ethanol currently produced is made from corn sugar and in order to reach the cost points required for such large production volumes, cheaper sources of sugar – such as cellulose and hemicellulose – will need to be used. The US Department of Energy has an active bioethanol program that supports a spectrum of technologies at various industries in the hopes that bioethanol can be made to be cost-competitive with fossil fuels. Biorefineries using these new approaches should be able to produce bioethanol and other useful products in a sustainable fashion.[9] It is anticipated that this industry can create hundreds of thousands of jobs worldwide – mostly in rural areas.[10]

Finally, while the use of carbohydrates as raw materials is likely to remain center stage for a long time, there is another very abundant source of carbon that is susceptible to biological transformations – namely methane. Associated with oil production, the world currently has 5,000 trillion cubic feet (16 trillion metric tons) of methane reserves, and much of what is currently produced is unused, simply burned on-site, producing carbon dioxide and little benefit. There are microbes that can live on methane and convert methane to biomass and other chemicals. A number of laboratories, including DuPont, are beginning to work with these microorganisms to determine how they might be used to produce materials from methane. While not a renewable resource per se, an effective, non-capital intensive method of utilizing methane for the production of chemicals and materials would be of considerable benefit to our economies and environment.

Biology as a Source of Inspiration for New Materials

Introductory chapters in biology textbooks often classify biological entities as either "structural" or "functional." Examples of "structural" materials range from single molecules through formed objects, including polymer filaments (e.g. elastins or silk), sheet structures (e.g. protein-β-sheets and lipid bilayers), as well as three-dimensional objects with considerable structural detail at a microscopic and macroscopic level (e.g. cartilage, bone, fingernails or hair). These materials, while having some biochemical activity, serve primarily to give structure and physical dimensions to biological organisms. In contrast, "functional" materials include simple molecules (e.g. enzymes), aggregate systems (e.g. transport systems and receptor-signaling systems) as well as complex macrostructures (e.g. flagella and organs) that provide and utilize various biochemical transformations of substances to perform their function. The functional materials are, by far, the most well studied to date. However, more recently, the research community has begun to focus on the more difficult task of understanding the molecular details of structural biomaterials and how to produce them commercially. While we are still in the very early stages of utilizing this information in commercial production, a few examples of this type of research may serve to indicate what the future might hold.

There is no lack of biological structures and materials with interesting or unusual attributes. Poets, painters, engineers and material scientists find natural materials a source of much inspiration. The business community recognizes the potential commercial value of many of these materials and structures, but manufacturing such materials at scale and at appropriate cost is extremely difficult. In the following section, I touch briefly on three examples of materials chosen to illustrate different physical phenomena, technical advances and economic considerations.

These are:

- Silk fiber, in which the weight and properties of the final material are determined by the biotechnology innovation.
- Proteins that increase the strength of composite inorganic material, when present at very low levels.
- Colorants that contain neither dyes nor pigments but create an illusion of intense color, based on a photonic effect.

Silk as a Structural Protein

Silk is one of the very first bio-based materials that biologists focused on. Silk is a protein filament of insect origin that has been used and appreciated for thousands of years due to its unique combination of tensile strength, stretch and recovery, fineness, easy dyeability, and its feel or "hand" as it is pulled across the skin. From a technical perspective it is a monofilament of a roughly pure protein that is relatively uncomplicated by other materials, tertiary structure or structural entanglements. The prospects of producing a silk that might have improved care characteristics, be less expensive, have an altered denier or cross-sectional shape, and comprise a continuous monofilament (rather than being the length spun by the insect), was commercially very attractive, and many laboratories and industrial companies began to explore how to do it. A number of companies, including DuPont, extensively studied insect silks for potential improvement, production and eventual commercialization. The company's experiences are perhaps illustrative of the rewards and frustrations of doing early research in bio-based materials.

As background, most fibers with elasticity, such as lycra, are composed of repeated sequences of "hard segments" and "soft segments" that vary along the length of the filament. The individual polymer microfibrils are bundled into larger fibers in a spinning or extrusion process. The individual polymer molecules interact in such a way as to allow "stretch" as well as "recovery." With this basic understanding, the research community postulated that a new generation of fibers was possible. This was based on the belief that the complete control of monomer sequence, as well as polymer length, afforded by genetically based synthesis was feasible and economically attractive. From a polymer perspective, the ability to specify precisely the sequence of twenty different amino acid monomers in a polyamide fiber promised an unparalleled ability to create materials with desired properties. The chemical nature and range of properties available in the twenty amino acids appeared to offer a unique and chemically rich set of building-blocks with which to fashion a new generation of improved "silks" whose range and sequence specificity could not be matched by chemical synthesis.[11]

The challenge for the protein engineer was to specifically exploit the inherent value of sequence control so as to get a resulting functionality within the fiber. Many silk-like protein polymers were designed and produced via fermentation, and a wide range of properties were observed. However, none possessed the particular combinations of strength and toughness that would allow for commercialization. Our understanding of DNA code sequence proved to be greatly superior to our understanding of how monomers combine to give macroscopic properties.

We also learned a great deal about the importance of the insect's internal spinning process. Within the insect, the protein polymer exists in at least three different phases: a viscous aqueous phase is stored in the gland just prior to spinning; a less viscous liquid crystalline phase immediately precedes fiber formation; and the fiber itself, emerging from the insect spinneret, has both crystalline and amorphous regions.[12] These aspects of the natural formation of silk proteins have large effects on the ultimate properties of the material produced and have proven to be extremely difficult to replicate.

As a result, the inherent promise of the technology has been largely unfulfilled to date, though many researchers in this area remain hopeful that advances in protein production and rapid screening methodologies coupled with enhanced fiber formation technologies will eventually allow significant, commercial production of designed structural proteins. Indeed, some companies have continued research and development in this area, producing novel silks in microbes, insects and the milk of animals. We are likely to see, initially, the use of these silks in low-volume/high-value markets, and are hopeful that learning from these ventures will prove helpful in developing further uses for designed structural proteins.

Proteins that Organize the Structure Around Them

Many marine organisms make complex structures largely from inorganic materials that exist in solution in ocean water. Diatoms and sponges make intricate three-dimensional structures of silicon oxides. All mollusks make a wide variety of shells from a largely calcium carbonate matrix, with structural integrity far exceeding normal mineral forms of calcium carbonate such as chalk or calcite. In both cases, the resultant structures, formed from repetitive crystallization of a basic pattern of protein and inorganic oxide, are astonishing.

A number of research groups have made remarkable progress in understanding the biological, chemical, physical and structural science behind these materials. In this review, we will highlight only the efforts of a few, but more extensive references are available.[13]

Dan Morse's group at the University of California at Santa Barbara has provided two examples of how a "touch of biotech," specifically a small amount of a protein, can dramatically alter a largely mineral structure. Their work is derived from studying two marine animals at the genetic, biochemical and nano scales.

The abalone is a mollusk native to the waters off the California coast. It has a shell that is a microlaminate composite of calcium carbonate crystals and that contains a low percentage of protein. The fracture-toughness of the resulting structure is three-thousand times greater than that of calcium carbonate alone. Although the structure contains only a small proportion of proteins, these proteins are directly responsible for the tremendous enhancement of strength of the material and the precise control of its unique nanostructure. As an example of the type of multidisciplinary research that is becoming increasingly common, it is enlightening to list the tools that this group has used for the study of these phenomena. The group combined state-of-the-art atomic force microscopy, X-ray diffraction, light microscopy and molecular biology to resolve the proteins, genes and molecular mechanisms responsible for this control.[14]

One of the most interesting findings of the group, from a strictly material perspective, was the discovery of the mechanism by which the marine animal controls the abrupt transition from one mineral form of calcium carbonate, calcite, to another mineral form, aragonite. This "genetic switch" must be activated, as the abalone animal produces both the bulk shell material as well as the flat pearl – or "mother of pearl" – which covers the internal surface of the shell. By purifying proteins from the calcite and aragonite crystals, the group was able to control atomic lattice orientation of calcite and aragonite crystals produced in vitro, producing both the thermodynamically favored form as well as the unfavored form simply by adding a small amount of the catalytic protein to the solution. These proteins make possible the synthesis of composite materials via precise control of the crystal structure at micron-scale dimensions.[15] Based on these results, this group has begun to experiment with potentially commercially useful materials such as polymetallic crystalline thin-films with useful semiconductor and magnetic properties.[16]

This protein-based control of inorganic precipitation from a solution is not limited to calcium carbonate. The same group has extended this research to include silica oxide materials naturally formed and accumulated by a marine sponge. In the sponge, glassy needles are produced, the exact function(s) of which remains unknown. Using similar techniques that draw on both biotechnology and nanoscale analytical and engineering tools, they discovered that proteins exist, and can be engineered, to produce a wide variety of silica forms and structures, including opal-like silica and high-performance silicone polymer networks, depending on substrate and protein provided. This group named the responsible family of proteins "silicateins." They expect that these proteins and precipitation phenomena may provide a new route to the environmentally benign synthesis of high-performance silicon-based materials.[17]

In all of these cases, the discovery of proteins that alter the inorganic environment around them, so as to produce structures of remarkable strength and diversity, was made possible by the increasing power of genetics, high-throughput screening, and the increasingly diminutive, yet robust, tools of science and engineering.

Proteins that Organize Light

The last example of inspirational biology is biology aimed at light refraction. The ability of regularly spaced biological structures, of differing refractive indices, to create color is remarkable. A wide variety of animals and plants have evolved structures that consist of regularly repeating units 100 to 1,000 nanometers apart that exploit these effects. These include not only the internal surface of the abalone shell listed above but also other marine organisms (e.g. sea mouse), microscopic life forms (e.g. iridoviruses in solution), and land creatures (e.g. insects). In all cases the color spectrum, intensity, hue, light fastness, and self-assembling nature of the surface coating would be expected to have significant commercial value if it could be successfully produced in a modern manufacturing process.

Interest has been expressed in these phenomena by a number of commercial sectors. Small quantities of such a unique material could easily find their way into materials used for security, tamper evidence and authentication. If larger quantities could be manufactured, uses in cosmetics, art and the aesthetics of small objects around us would easily be conceivable. At still larger volumes and reduced costs, the manufacturers of paints, clothing and automobiles have all expressed interest.

All photonic effects are caused by the manner in which light is reflected off structures. A photonic effect results when a material is constructed in such a way that light reflected from one object interferes with light reflected from another object. The extent of constructive or destructive interference, and the color and intensity of the resultant hue, depends on the phase difference between the two reflections. Variables that affect this include total thickness of the structure, refractive indices of the materials that make it up, as well as incident angle and wavelength of the light illuminating the structure. Nature has evolved hundreds of photonic structures representing constructive and destructive light patterns in the visible as well as the ultraviolet and near infrared regions of the spectrum. Many of these would have considerable economic value if they could be repeated on an industrial scale.

A large number of insects exhibit photonic coloration and have been extensively studied. One such insect is a butterfly native to South America. This butterfly, the blue morpho (*Morpho rhetenor*), is particularly well studied by a group at the University of Exeter.[18] The photonic effect is produced by complex three-dimensional structures at the surface of a roughly planar scale of the wing. Unfortunately, although powerful microanalytical tools have been used to deconvolute both the mechanism and the relevant structures, the use of molecular biology tools has proven limited in this case. Self-assembling microstructures of sufficient size and placed at appropriate spacing for photonic scattering to occur is proving to be a challenging feat from a biological perspective. Paralleling the case of silk mentioned above, an inspirational insect structure is proving difficult to repeat at a manufacturing scale.

Despite this, the bio-inspired photonic effect has been produced at one appropriate scale to be seen in the visible range in a new film from 3M.[19] I mention this to make an additional point. Often when a biological molecule or process inspires the biological community to follow a particularly promising lead, a concomitant advance in the chemical and engineering community indicates a way to provide part of the benefit sought by investigating the biological approach. Chemical problems also inspire biologists to seek their own approaches to finding solutions. This type of multidisciplinary-inspired problem solving is gradually becoming the norm, and I believe it will increasingly be a hallmark of future advances in materials sciences.

One illustrative example of this encroachment of biotech into the materials discovery world is in the "phage-display" efforts of researchers such as Angela Belcher, now at MIT. Phages are viruses that infect bacteria and are used by genetic engineers to introduce genes into bacterial cells. The phage-display technique uses a library of phage particles, each of which has displayed on an external surface a sequence of seven to twenty amino acids. In this manner, upwards of 10^8 different sequences can be screened in a single experiment. The phage particles are placed in contact with the surface in question and allowed to bind. Particles that do not bind are rinsed away. Bound particles are allowed to multiply and then the process is repeated a number of times, with increasing stringency of rinsing (including additional increments of surfactant). Phage that sticks to the surface, even after vigorous attempts at removal, must have reasonable binding affinity, the specificity of which is checked. Phage-display technologies combine many aspects of classical and Darwinian biology, from generating diversity, through microbial and viral ecology, survival of the fittest and natural selection. The resulting system allows one of the most high-throughput discovery systems for polypeptides ever conceived.[20]

Dr Belcher's lab used this system specifically to discover novel attachment systems (peptides) with binding specificity for a semiconductor component that would enable directed nanocrystal assembly.[21] The results of this work show the remarkable specificity of a peptide sequence for a specific face GaAs(100), over the (111)A (gallium-terminated) or (111)B (arsenic-terminated) face of GaAs. They further extended this peptide recognition and specificity of inorganic crystals to other substrates, including GaN, ZnS, CdS, Fe_3O_4 and $CaCO_3$, giving rise to the possibility that a new generation of fabricated electronic devices and structures may be produced outside the traditional fabrication facility.

This group – and others since then – have shown the power of using phage-display libraries to identify, develop and amplify binding between organic peptide sequences and inorganic semiconductor substrates. Although the ambitions to build an electronic device in solution may still be a long way off, the development of specific adhesives and self-assembling materials is being attempted in a number of high-value marketplaces, including medical care, sensors, electronics and coatings.

Biotechnology's Role in a Sustainable and Renewable Future

These are just a few examples of the uses of biotechnology for industrial materials productions. There are many more, and I encourage those interested to consult some excellent monographs on the topic. However, I hope these instances serve to illustrate the optimism with which one can contemplate the future of biotechnology as applied to materials production.

The benefits of using biotechnology to produce industrial chemicals and materials have been most keenly felt in the production of pharmaceuticals, specialty chemicals and food ingredients. Many of these have used enzyme-catalyzed reactions as part of a larger production process, and some of these processes are truly large scale, as in the production of high-fructose corn syrup for sweetening foods (8 million metric tons/year) and the production of acrylamide (20,000 tons/year) for materials uses.

Examples of more complex transformations of raw materials into industrial products by fermentation are perhaps less well known, and in the past have been primarily associated with food-related molecules – vitamins, amino acids and so forth. However, the very large-scale production of ethanol as an alternative fuel, and the recent examples of polylactic acid and propanediol, represent major, large-scale uses for this technology.

Where biotechnological approaches have been used, the processes tend to use less energy and lower amounts of raw materials and capital investment, as well as generating less waste than the corresponding chemical approaches. However, when there is a choice between a chemical and biochemical route of production, the usual parameters of operating cost, capital investment and environmental burden all play a role in making the choice of which one to use. When contemplating a biotechnological approach to manufacture, one must be cognizant of the huge capital investment in chemical manufacturing facilities that has been made in this past century, and the many years of experience in chemistry and engineering that have produced world-scale facilities that efficiently utilize cheap raw materials derived from petroleum. These factors create a highly competitive environment against which any new technology will be judged. Well-capitalized, highly automated chemical technology is the incumbent technology – its often-depreciated facilities are not easily dismissed. Often, on a simple cost basis, a biotechnological route – despite numerous other advantages – will just not be competitive. In other cases, one can find some particular steps in an existing industrial process that would benefit from a biotechnological approach. The artful practitioner would do well to take great care in determining how to proceed.

Still, there are a number of very large forces at work that, to my mind, will ensure the adoption of biotechnology in materials production. I have already mentioned the significant environmental benefits that can accompany biotechnological approaches. In addition, enzymes and microorganisms are well adapted to using precisely the products that come from renewable production of materials; namely, sugars and other oxidized materials derived from agriculture and biomass. At the same time, advances in biotechnological methodologies allow us to better utilize these agricultural commodities to produce valuable products. This is causing companies that have traditionally been in very different markets to find new bases for collaboration or competition. Biotechnology is bringing together companies with materials expertise and companies with expertise in the production, transport and use of agricultural raw materials. Examples include Cargill-Dow for the production of polylactic acid, and DuPont and Tate & Lyle for the production of propanediol. It will be very interesting to watch how this plays out.

Having said that, agricultural commodities have their first use as food, and this use will always predominate over materials uses. To my mind, a real breakthrough will occur when we learn to utilize the waste products of agriculture – the stalks, straws and bagasse. A number of companies, supported in part by government funds, are seeking just that. Successful use of these types of raw materials will require a significant marriage of chemical, engineering and biotechnologies but, if successful, promises to reduce raw materials costs for biotechnological transformations by a factor of two. This would have an enormous impact on the variety of chemicals and materials that could be economically produced by biotechnology.

As just mentioned, the elaboration of new technologies often requires governmental support to help "prime the pump." This is occurring, both in terms of the usual support of basic research in universities and, increasingly, through the use of targeted funds to encourage consortia of companies and universities to embark on highly risky, expensive development efforts to learn how to utilize renewable resources. Governments and society are compelled by the vision of a renewable future in which there is a better balance between production and consumption of raw materials and between the amount of carbon dioxide consumed and liberated. Biotechnology is particularly well suited to aid in this move toward a more sustainable future.

Companies seeking to utilize biotechnology will need to work with the government, not just as recipients of funds, but also with governmental regulators to ensure that appropriate regulations exist to safeguard the environment as biotechnological production increases. Biotechnology is not zero waste, though the major waste products such as spent enzymes and biomass tend to be biodegradable or useful as feed additives, and are generally fairly easy to deal with. However, as production increases, care will need to be taken to ensure that this remains the case.

In addition to these large-scale "traditional" industrial uses, some recent biotechnological advances are mentioned that rely not on the manufacturing capacity of biotechnological approaches but rather on the use of biotechnology to develop brand new materials with new properties that can be used in dramatically lower amounts than incumbent materials. These aspects of biotechnology are much more recent and, to my mind, do not yet form a coherent pattern of development that allows one to make precise predictions regarding their commercial utility. However, in addition to sustainable production of large-volume materials, substitution of lower amounts of materials to achieve the same ends is also a worthy goal of a sustainable future. Such materials will find use in electronics, coatings, sensors and biomedical applications, and will start to appear over the next ten years. They will play a role in the ongoing "dematerialization" of society and will occur over the same time period that we learn to utilize renewable resources for production of bulk chemicals and materials.

In my view, we are at a very interesting point in time. Biotechnology has clearly shown itself to be "ready for prime time," and there are numerous new approaches in the research and development pipeline that will generate new commercial realities in the near future. One of the most powerful new forces at work is the generation of truly multidisciplinary approaches to problem solving. While the value of such approaches has long been known and discussed, it has been my experience that in the area of biology, it is truly starting to happen. Indeed, many of the tools of modern biotechnology – such as high-throughput DNA sequencing – derive from precisely this kind of interdisciplinary approach. Scientists and engineers with very different background, vocabularies and outlooks are beginning to collaborate across a large technological space. This is exceedingly difficult work, and those who do it best will be the winners. Some examples have been provided of this type of work, where physics, materials science and biology all play key roles in determining the outcome of experimentation, and there will be many more in the future.

We will have examples of biotech-centric approaches in "tour de force" fermentation approaches to producing industrial materials. We will also have examples of materials-centric approaches that use just a minimum – albeit critical – amount of biological input. We will continue the pattern of biotechnologies initially producing higher-value/lower-volume materials, and over time the increased application of multidisciplinary approaches will enable the production of larger-volume/lower-value materials. In future, we will learn how to take advantage of the lessons of biology in producing exquisitely designed new materials based on biological principles. Most of these developments will result in processes that are cheaper, more environmentally sound and, hence, more sustainable.

The rate of progress in the use of biotechnology to produce materials is such that the elaboration of a truly multidisciplinary approach to solving a number of our material problems is upon us. Given this rate of development, twenty years from now biotechnology will have taken its place as another set of skills and approaches used by the materials industry, and it will no longer be particularly relevant or insightful to talk about "Biotechnology and our Material Future." When this occurs, society will have taken an enormous step towards ensuring a sustainable future.

9

Biotechnology and the Agricultural Industry of the Future

Ray A. Goldberg

Advances in biotechnology are transforming the entire agricultural and food system of the world. In highlighting these changes, this chapter will address consumers' perceptions of the positive and negative aspects of genetically modified crops and animals, and also discuss the impact of biotechnology on the Third World and its relationship to the global economy.

The role of agricultural multinationals, government research organizations, foundations and international institutions are becoming increasingly important as the research and development of new resistant, high yield and high value-added crops require considerable amounts of funding. Major agricultural companies are focused on gene discovery, gene mapping and gene sequencing of humans, plants and livestock and of their pathogens, and are investing heavily in the new biotechnologies. In fact, the supply side in the world food system now begins with participants who generate and supply fundamental biochemical and genetic information. At the same time, the traditional role of farmers is continuing to shrink. In 1950, the value added by farmers to global agribusiness was 32 percent. By 2028, farmers will add only an estimated 10 percent of value. Food processing and distribution accounted for 50 percent of the added value in 1950 and will account for over 80 percent of value in 2028. Concurrently, the responsibilities of the farmer are being extended to cover not only the production of food, feed, fiber and energy products but also the production of pharmaceuticals. Farmers have the additional responsibility of maintaining land and water resources in an environmentally sound manner. However, despite the fall in the value-added share on a global basis, the farmer has gained greater importance in the global food and "agriceutical" system.

The newly discovered ability to map the genome of humans, plants and animals so as to improve health, nutrition, safety, the environment and the global economy is colliding with the mistrust of the science by some groups. Concerns are being voiced regarding the safety of genetically modified crops as well as the use of agricultural biotechnology in developing countries. Many consumers remain skeptical about bioscience advances, including genetically modified foods. However, tests have not validated the safety concerns regarding commercial GM crops. Accordingly, the US Food and Drug Administration (FDA) has decided that no toxicological testing of genetically modified foods will be required.

The safety of food supply is still very much on the agenda of Third World countries. Banning genetically modified foods could be disastrous for developing countries. Biotechnology can help to produce higher yields on less land, thus generating crops beyond mere subsistence level. With conventional farming methods, the additional land required would have necessitated cutting down significant amounts of forests. The so-called "Green Revolution" to alleviate world hunger rests to a large extent on genetic modifications that enable crops to resist extreme climatic and soil conditions.

Simultaneously, the developing world faces pressure from some markets (such as the European Union) to refrain from using the new science, as their products may not be welcomed in these markets. In essence, the stand being taken is: "If you want to sell to us, don't use genetically modified organisms (GMO) or products."

In the context of the genetic revolution, the issues impacting on the global food system are not just those of productivity but also of the environment, world trade, nutrition, food safety, health, and the political stability of nations and regions. Access to technology and intellectual property safeguards are also major concerns. If consensus cannot be reached on these critical issues, economic development in both the developing and developed world will be threatened and the burden of this lack of trust will fall heavily on the poor small-scale farmer and the low-income consumer.

The Transforming Impact of the Genetic Revolution

The new genetic research revolution will change the global economy and society more dramatically than any other single event in the history of humankind. It will change our perceptions regarding firms, institutions and industries, and affect our position and trade relations between countries, customers, suppliers and investments as well as the relationship between private, public and non-profit institutions and society.

The genetic revolution is leading to an industrial convergence of the food, health, medicine, fiber and energy businesses. The mapping of the male and female human genome, as well as the genomes of the fruit fly, the mustard seed, the mouse and the rice grain have taken place. As organisms evolve, they usually retain many of their old genes, which means that most life forms such as plants, animals and humans share similar genetic structures and immune systems. The overlapping of these structures is leading to new discoveries that will improve the health, nutrition, safety, environment and economic development of society. On the other hand, for the food systems sector of this new life science industry, we have some particular concerns: a worldwide agricultural economic depression; a mistrust of the science by some consumers; anxiety regarding the choices available to consumers; the intellectual property rights of the public, private and non-profit sectors; and the appropriate rules, regulations, labeling and testing necessary to protect consumers and the environment from potential and unforeseen risks.

In view of such uncertainties, I approached the Provost of Harvard University eight years ago to form a university-wide committee with members from the Business School, the School of Government, the School of Public Health and the Medical School to work with the value-added food chain comprising consumer groups, producers, government leaders, business leaders and technology leaders. The objective was to ensure that this most important economic, political and social revolution in the history of humankind would serve and be responsive to society and be dedicated to improving our food, feed, energy and pharmaceutical system. This ought to be done in a manner that is safe, environmentally sound, and improves the economic development of all sectors of the global economy.

One of the consumer members of the Private and Public, Scientific and Consumer Policy Group (called PAPSAC) has made a major contribution to a recent National Academy of Science report. The National Academy of Science's Committee on Genetically Modified Pest-Protected Plants reviewed the scientific data on potential health and environmental risks and the use of these data in the regulatory process. The committee recommended that priority be given to developing improved methods for identifying potential allergens, specifically focusing on new tests relevant to the human immune system. It wanted updated guidelines to coordinate plant biotechnology regulation activities of the Environmental Protection Agency (EPA), the US Department of Agriculture (USDA) and the Food and Drug Administration (FDA). The committee also recommended that agencies monitor the ecological impact of pest-protected crops on a long-term basis to ensure the detection of problems that may not have been predicted from tests conducted during the registration and approval process. In other words, any new technology requires constant and long-term monitoring by independent third parties and the provision of escape hatches in case people make mistakes in judgment or operating procedures.

The Implications of the Life Sciences Revolution

The first implication of the life science revolution is to redefine the boundaries of the agribusiness system to one that has become an "agriceutical" system.

Figure 9.1
Agribusiness Systems Approach

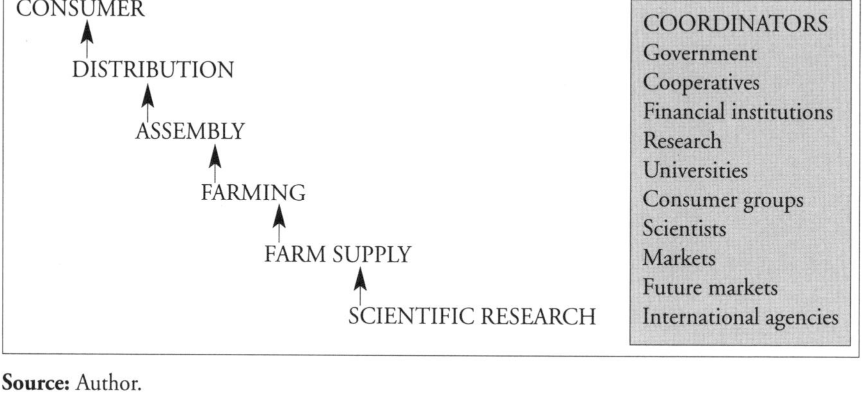

Source: Author.

Figure 9.2
The Agribusiness–Agriceutical Transition

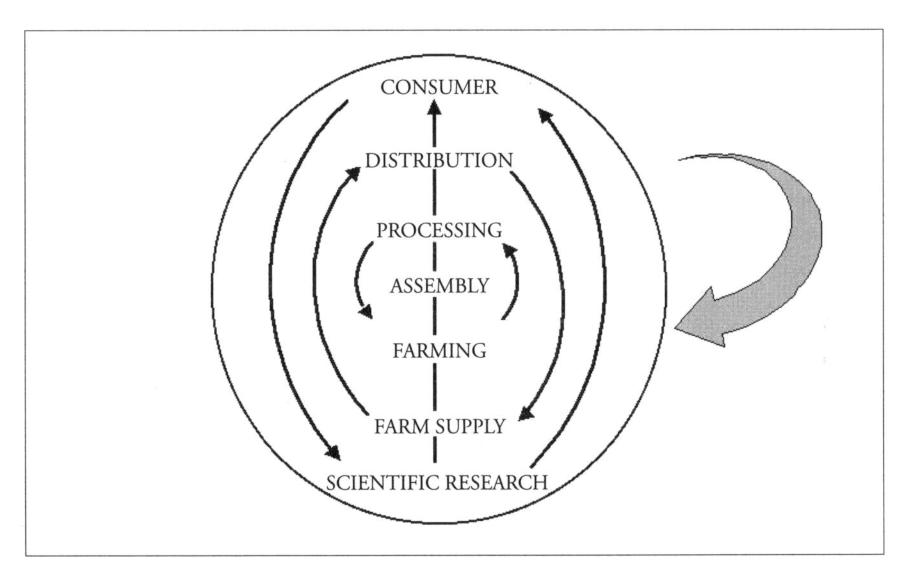

Source: Author.

Figure 9.3
The Global Agriceutical System

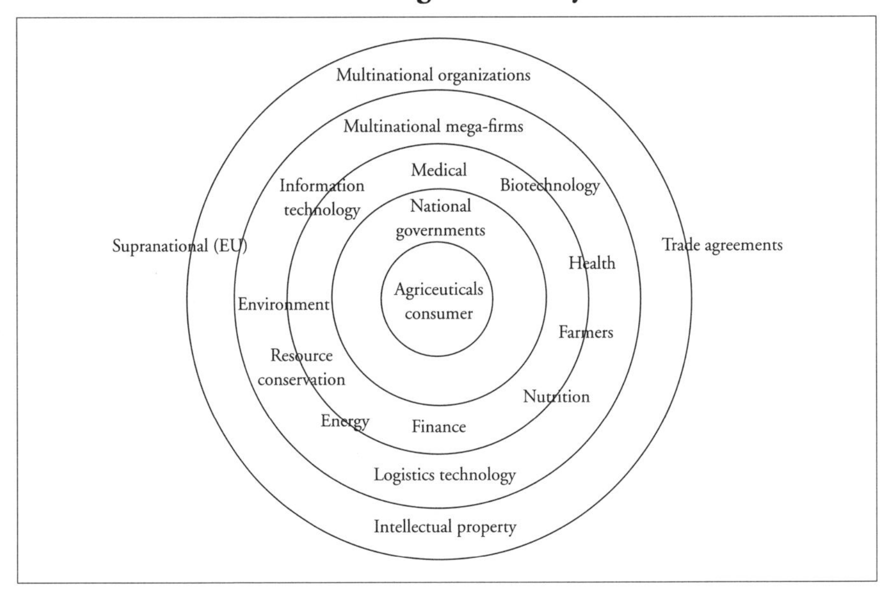

Source: Author.

From Figures 9.1, 9.2 and 9.3 and Table 9.1, it may be noted that the traditional agribusiness system without the pharmaceutical, health and life science segments will be an eight trillion dollar global industry by the year 2028. The farming sector value added will have shrunk from 32 percent in 1950 to 10 percent in 2028. This trend does not mean that farmers are no longer important – they are essential change makers of the food system and will be creating identity-preserved measures and traceable products, as well as providing the consumer with more choices. Whereas food processing and distribution accounted for half of the value added during the 1950s, it will account for over 80 percent of value added in 2028.

Table 9.1
The Value Added in Global Agribusiness: Marked Shift Towards End Product Between 1950, 2000 and 2028 (billions of US dollars)

	1950			*2000*			*2028*		
	Market	*Value added*	*%*	*Market*	*Value added*	*%*	*Market*	*Value added*	*%*
Food supplies	$44	$44	18%	$520	$520	13%	$720	$720	9%
Farming	125	81	32	1,120	600	15	1,520	800	10
Food processing and distribution	250	125	50	4,000	2,880	72	8,000	6,480	81
Total	$419	$250	100%	$5,640	$4,000	100%	$10,240	$8,000	100%

Source: Author.

The addition of life science participants in the new agriceutical system will increase total value added in 2028 to over fifteen trillion dollars and the farmers' share will shrink even further to 7 percent (See Table 9.2). This new agriceutical system is much more complex. The immune system of plants, animals and humans interact with one another and their common environment. The institutions that are created to deal with this new global complex are also global in nature, such as the Codex Alimentarius.

Table 9.2
Global Agriceutical System 2028 (billions of US dollars)

	Value added	*Percentage*
Life science companies and farm suppliers	1,400	9
Farming	1,000	7
Food and health processing and distribution	12,960	84
	US$15,360	100

Source: Author's estimates based on data from the US Department of Agriculture and US Department of Commerce.

The second implication is that the revolution leads to changing perceptions about the means to measure the potential benefits to the participants in the system and the society that they serve. How do we share the benefits of cost reductions? How do we share or even measure the value of improving the health and longevity of our consumers? Traditionally, the industry has tried to give a majority of the benefits at the production level to the producer. Similarly, through the operation of competitive forces, the major proportion of cost savings and product improvements have been passed on to the consumer. In recasting the industry, vast sums have been paid for those companies that have access to germplasm or gene sequencing. These investments have meant that firms have banked their whole future on the potential of life science. It is a recognition of the fact that, at the end of the day, these are the new technology firms of life science. This new system is large in terms of having 50 percent of the global labor force, 50 percent of global assets and over 40 percent of consumer expenditures. At the same time, it is small in the sense that consolidation, contractual integration and partnering have reduced the number of decision points in the system.

The third implication of the life science revolution is the difficulty of finding neutral third parties to evaluate and monitor the impact of new technology use and its effect on the environment, health and safety of consumers. So many features of the technology provide us with new ways to reduce pollution and to measure more precisely the impact of alternative policies and procedures on the environment. Some industries, such as the wood and pulp industry, have encouraged third-party evaluations on the environmental impact of their policies such as was done by the

International Institute for Environment and Development. It is imperative that all industries develop such policies and procedures to respond to justifiable concerns of environmental and consumer activist groups. The model of the pulp and wood industry is an excellent one.

The fourth implication of the life science revolution is the potential to misuse it as a potential trade restriction in the form of a "precautionary principle" or "fourth hurdle."

The fifth implication is that the technology can be abused – become a weapon – or confront the common ethical values of all societies.

The sixth implication is that the patent activity for genes, plants and animals will encourage networks, partnerships and cross-licensing that will give all firms access to and some firms ownership of the valuable intellectual capital currently being developed. These networks will cross old industry borders. With thousands of new compounds and procedures being discovered yearly, a company in one industry may uncover and patent a solution to a problem that a company in a very different industry has been working on for decades.

The seventh implication is the recognition that firms involved in spending huge sums on research and development need intellectual property rights to obtain fair returns on their investments. Simultaneously, the issue of access to this technology has to be addressed. One approach to this issue may be seen in a recent agreement between the National Institutes of Health (NIH) and E.I. DuPont de Nemours and Co. of Wilmington, Delaware. As of January 1, 2000, the agreement allows NIH-funded scientists doing non-commercial research to use patented transgenic animals (such as the Oncomouse) without obtaining the specific approval of DuPont. Even more recently, the Monsanto Company has released the rice genome, which will enhance the research of the World Bank and Rockefeller Foundation Seed Center throughout the world. Poor nations must be able to tap into the benefits for bioengineering. Dr Gordon Conway, President of the Rockefeller Foundation, has observed that without access to bioengineering, the Third World would be unable to maintain an expanding population.[1] In the past, seed companies and pharmaceutical companies have shared new technologies with public and nonprofit institutions. This sharing will continue and many new discoveries will come from the public and nonprofit sector as well as the private sector.

The eighth implication is that the new technology will lead to identity-preserved crops and livestock providing traceable production and distribution systems, which in turn will encourage the creation and use of new institutions to provide common global labeling, common definitions of organic or other physical attributes, and a widely accepted traceability system with appropriate fines and disciplines for participants that abuse the system. Codex Alimentarius is an excellent model for the future.[2]

The ninth implication is the impact that the life science has on our perceptions of what we do and how we do it. The function of all the participants in the system has changed, as indicated in Figure 9.4.

Figure 9.4
Functions Redefined

(1) Farmer has become a technology and resource manager
(2) Seed feed and machinery supplier has become a life science company
(3) Commodity handler and processor has become an ingredient supplier
(4) Brand food and beverage supplier has become a nutrition and taste inventor
(5) Distributor has become a consumer advocate and knowledge network supplier
(6) Government has to be multinational and multi-department
(7) Food system is a model for others in responding to the economic development needs of society

Source: Author.

The farm supply firm has become a life science company. The large- and small-scale farmer has become a custodian of land and water resources and an applier of genetic technology (technology that reduces the cost of food, fiber, fuel, and plant- and animal-produced pharmaceuticals) and adds value in terms of health and nutrition in a manner that reduces pollution and provides for long-term sustainability. The farmer is also expected to manage the process in a way that safeguards the diversity of germplasm and wildlife. The small-scale producer who receives most of his or her income from non-farm activities will become the keeper of the rural lifestyle and rural community so important to the maintenance of our respective value systems. In return for these improvements and appropriate use of technology, farmers will become contractual partners to the rest of the agriceutical system, with appropriate income safety nets to accommodate the global price and production volatilities that exist at the farm level.

The assembler acts as a handler of identity-preserved products who ensures origins and safety at all levels of the food chain, or, in other words, traceability. The processor is a developer of branded and own-label food products, and of branded and generic pharmaceutical products, and provides not only caloric content but also health and nutritional alternatives. The production of pharmaceuticals in herds of goats or in field crops provides low-cost alternatives to heavy investments in plant and equipment. However, this mode of production also requires safeguards to avoid such animals entering the general food supply chain. The distributor becomes the partner of the consumer, now working in partnership with hospitals and health systems to provide unique foods for people to control and manage disease, and to enhance the health and nutrition of the general population.

All of these activities require close working relationships between the private and public decision-making systems; in essence, the companies and the regulators. It also requires the involvement of consumer groups. Furthermore, national decision making will require a sophisticated activity-based accounting system that will both measure the progress and effectiveness of the separate segments of the agricultural system and indicate the approximate level of involvement for each firm in the system. The agriceutical system is here to stay and we must understand how best to organize and utilize it. Genetics in the past has represented 56 percent of the total yield gain. Genetics in the future will continue to contribute to an improved diet for humans and animals and will also account for over 50 percent of the increase in our longevity. It is necessary to reach out to consumer groups and others in order to continue to develop this critical technology and to improve our human and natural resources in a fair and equitable manner. We have the obligation to renew our resources, our institutions and ourselves.

The tenth implication of the biotechnology revolution is how US consumers view this revolution. From Tables 9.3 and 9.4, it is clear that issues of food safety, ingredient labeling and pesticide labeling are seen as more important than GMO labeling.

Table 9.3
What US Consumers Want on Food Labels

(1)	Food safety	24%
(2)	Ingredients	20%
(3)	Fat-caloric-sugar content	14%
(4)	Grown with pesticides	12%
(5)	Genetically modified	4%

Source: Alliance for Better Foods.

Table 9.4
Other Activities of US Consumers

(1)	71% aware of bio-engineered foods
(2)	17%–24% concerned about bio-engineered foods

Source: Alliance for Better Foods.

From Table 9.5, it may be observed that Europeans put GMO labeling versus organic labeling above other food safety concerns.

Table 9.5
Consumer Survey Results on Food Safety Issues

Issue	*Percentage*
Food safety	68
Absence of pesticides	54
Absence of hormones	56
"Organic" without chemical pesticides	81
Biotechnology should be labeled	86
"GM-free" label	77

Source: Eurobarometre 49.

From Figure 9.5, it is evident that the global adoption of biotech crops has been rapid. In 2003, the International Service for the Acquisition of Agribiotech Applications (ISAAA) reported that 6 million farmers in 16 countries planted genetically modified seeds on 145 million acres – an increase of over 10 percent over 2001. The report indicated that the value of bioengineered crops would reach US$5 billion by 2005 compared with US$4.25 billion in 2002.[3]

Figure 9.5
Worldwide Planted Acres of Biotech Crops (million acres)

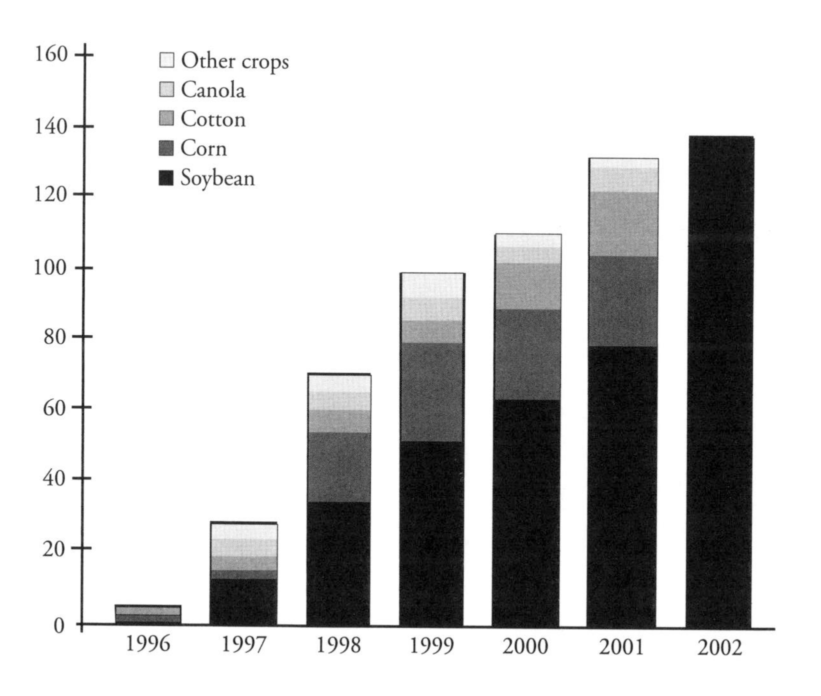

Notes: Global adoption of biotech crops has been rapid. It is estimated that 5.5 million farmers planted biotech crops in 2001 (more than 75 percent of them from resource-poor nations)

- Biotech planted area has grown > 30 fold since 1996
- The number of countries planting biotech crops reached 13
- 2001 acreage grew 18%
- Continued growth expected in 2002
- Biotech acreage share:

	Global	US
Soybean	46%	74%
Cotton	20%	76%
Canola	11%	67%
Corn	7%	27%

Source: 1999, 2000, 2001, 2002, US Department of Agriculture and Monsanto estimates.

Conclusions

The key points that have been discussed in this chapter may be briefly summarized below:

- Genetically modified organisms (GMOs) have been with us since Barbara McClintock won her Nobel Prize for jumping genes.
- All plants, animals and humans have been genetically modified over time.
- Environmental issues such as Prodigene – with pharmaceuticals in plants and the pollen drift in Star Link – indicate there really are no 100 percent GMO-free products. What is the level of tolerance or criteria required to state that something is GMO-free?
- Much of the criticism of the science is really a criticism of the firms that are leaders in the science. There is a question of trust involved – trust of the private sector, the public sector and the scientists.
- While ensuring protection for intellectual property, the question of who has access to the technology remains an important issue.
- The role of the Global Conservation Trust to protect biodiversity and germplasm in the future is a critical one.[4]
- Consumers want and deserve a choice. Labeling will occur under the global leadership of Codex Alimentarius.
- The genetic revolution is a global revolution, not a domestic one.
- The developing world needs the technology as much or more than the developed world to adjust to harsh climates and small-scale agricultural needs.
- The National Academies of Science in both developed and developing countries should support the science together with appropriate safeguards.
- Contractual integration will lead to identity-preserved crops for both GMO and non-GMO crops.
- Codex Alimentarius is a model institution for the new agriceutical system and its leader, Alan Randall, is a model administrator.
- A new breed of private and public leader is needed in order to move forward and overcome the distrust of the past.
- Finally, the private, public, farm cooperative, nonprofit and consumer sectors have to work together in the future to maximize the opportunities of the genetic revolution and to develop rules, regulations,

procedures and monitoring to protect society from any unforeseen problems.

Section 5

Biotechnology and Health

10

Biotechnology and the Future of Medicine

*Gregory Stock**

In order to see how advances in biotechnology will reshape the practice of medicine, it is necessary to step back and take a look at the broad shifts being brought by our growing understanding of the underlying workings of life. The biggest challenges will come not from better medicines, improved diagnostics and more effective treatments for disease. These fit comfortably into the existing medical framework of healing the sick, reducing suffering and doing no harm. The real tests will come from developments that threaten the fundamental structure of medicine by blurring the lines between therapy and enhancement, prevention and treatment, and need and desire; by altering the relationships between patient and physician; by shifting our underlying perceptions of ourselves; and by changing even the basic trajectory of human life. Moreover, these larger challenges are precisely what lie before us as we continue to unravel the inner processes of life.

Expectations of such transformative developments are not new. In 1780, Benjamin Franklin, in a letter to his friend, the great chemist Joseph Priestley, commented:

* The author explores the topics raised in this chapter more fully in his book *Redesigning Humans: Our Inevitable Genetic Future* (Boston, MA: Houghton Mifflin, 2002).

> The rapid progress true science makes occasions my regretting sometimes that I was born so soon. It is impossible to imagine the heights to which may be carried, in a thousand years, the power of man over matter.

And now, a mere 225 years later, the pace of scientific advance is so rapid and the immediate possibilities so amazing that Franklin's observation seems tame.

The genomics revolution is emblematic of a host of new technologies in molecular biology and biomedical research that are poised to carry us quite literally to destinations of new imagination. Together, they will transform medicine and humanity itself. Before we explore these extraordinary possibilities, let us look at some of the dilemmas that the infusion of high technology into medicine will create in the next few decades.

As the capabilities of biomedicine continue to advance, we will gain ever more influence in realms that were once beyond our reach. Not everything is possible, of course, but new diagnostic and therapeutic capabilities that we only glimpse today will become commonplace tomorrow, and their arrival will bring difficult new knowledge, hard choices and painful tradeoffs. Do we really want to learn from our genes that we are at high risk of Alzheimer's or heart disease? Are we prepared to decide whether to withdraw life support from a fading, decrepit parent? How many millions in health insurance should we spend on round-the-clock nursing care and incubators to save a two-pound, 26-week-old premature baby? Do we want to block the use of human embryos to grow tissue for sick adults? What criteria should we apply to choose among those vying for a healthy heart that is available for transplantation? Will we seek to impose an age limit beyond which a woman (or man) will be barred from having a child using advanced reproductive technologies? Should parents be able to choose the sex or even the temperament and personality of their children?

The core questions here are simple ones: Who pays? Who decides? Who has access to health information? What procedures are acceptable? What risks are too high? What role should money and advertising play in healthcare? What sway should religion and government have over individual medical choices? The answers are not so simple.

No matter how we move ahead, we will have successes and failures. Some people will get hurt; others will be saved. While some will abuse these technologies, others will use them wisely. And, of course, some will claim we are going too fast and must slow down, while others will insist we are going too slowly. There will be no general consensus about how to proceed or even about whether these developments are a wonder or a horror.

Impact of the Genomics Revolution

In order to understand the direction in which medicine is headed, it is critical to understand why today's genomics revolution is so profound. In the summer of 2000 when Craig Venter and Francis Collins announced the rough sequence of the human genome[1] – the full complement of our genes and other genetic material – hyperbolic metaphor abounded. As scientists and pundits groped for ways of expressing the immensity of the occurrence, they often fell back on religious metaphors, even though many of them are not even religious. Article after article described how we were deciphering "The Code of Codes," reading "The Book of Life," or finding the "Holy Grail of Biology." This exuberance was reminiscent of 1969, when Neil Armstrong walked on the moon, and we were about to zoom off towards the stars. However, 2001 is behind us now, and not only is HAL – the advanced computer intelligence in Stanley Kubrick's *Space Odyssey: 2001* – nowhere to be seen, we have yet to embark on an odyssey to our own moon, much less the moons of Jupiter.

It seems reasonable to wonder whether thirty years from now, when the present hoopla about genes is long past, we, or perhaps our children, will look back in the same way and dismiss today's hopes about genomics as an equally wild dream, an exploration that ultimately led nowhere and did little to alter medicine or change the human condition.

Such a scenario is unlikely. Unlike space, genetics is at the core of our being. As we learn to modify our genes and other aspects of our biology, we are learning to modify ourselves. The capacity of technology to alter our lives profoundly is already clear from the extent to which we have employed it to reshape the world around us. The cavernous structures of steel, glass and concrete in New York, Hong Kong or Abu Dhabi are a far cry from the stomping grounds of our Pleistocene ancestors, and now our technology has achieved such power and precision that we can turn it back upon ourselves. The ultimate consequences are clear. Eventually, we will reshape ourselves as profoundly as we have already reshaped the world around us.

The revolution now underway in genomics and the rest of molecular biology will do more than revolutionize medicine, it will alter huge swaths of our economy, change how we have children, modify the way we manage our emotions, and even extend the human lifespan. So profound are these developments that one day they will force us to examine the very question of what it means to be human, and that day may come sooner than we imagine.

Two unprecedented developments are underway. The first is the Silicon Revolution. Among other things, this has spawned the emergence of the Internet and the diverse communications and computational devices that are so changing our world. In essence, we are breathing into inert silicon – the very sand at our feet – a complexity that rivals life itself. This is a momentous breakthrough not just in the history of humanity but also in the 3.5-billion-year history of life itself. We can barely glimpse the dimensions of change that will attend this development, but life on this planet will never be the same.

The second development, the Biotech Revolution, is a child of the first. It will impact us even more deeply, because it will give us increasing control over our biology and our evolutionary future. Despite the predictions of techno-enthusiasts about future medical procedures that turn us into Cyborgian fusions of flesh and silicon, biotechnology and its manipulations of our biology and biochemistry will matter most to us, our children, and their children.

Much has already been accomplished. Virtually the entire complement of genes that humans possess has been sequenced. The several million most common variants of these genes have been identified – the subtle genetic distinctions that make each of us biologically unique. We have developed inexpensive DNA-chip technology to read our genetic constitutions and record patterns of gene expression. And we have developed computer technology and bioinformatic methods capable of deciphering the masses of genetic data that our sequencing efforts are producing. In short, we know our genes, we have libraries cataloguing their types, we have tools to read them broadly, and we have the computational resources to make sense of this tsunami of genetic information about to wash over humanity.

The pace of the technology is mind-boggling. In 1998, a DNA chip to read perhaps 20,000 snips of DNA cost about US$8,000. Today, a chip to read more than 200,000 DNA segments costs less than US$200. Craig Venter and others predict that by 2007 it will cost less than US$1,000 to completely sequence an individual's entire genome, which means that for a few hundred dollars, we could look at the entire set of SNPs (single nucleotide polymorphisms) that reveal the idiosyncrasies of our genetic constitutions. In other words, we will soon have access to the details of our individual genetic constitutions and begin to read the story written therein.[2]

Important efforts in this field include that of Decode Genetics, where Kari Stefansson is sifting the genealogical and health records of the entire population of Iceland and cross-correlating the records with genetic information to hunt for disease genes. Yet another effort is in Sardinia, where the genetics of centenarians in several small villages inhabited by exceptionally high numbers of these long-lived individuals are being explored. Also noteworthy is the work of EGeen in Estonia, where the health records of that population are to be the basis for their nation becoming a laboratory for medical research and pharmaceutical trials.

Notwithstanding the hyperbole and excitement about genomics, little in medical science seems to have changed since the sequencing of the human genome. Heart disease and cancer are still the number one killers. We still age and grow frail and we still suffer from genetic diseases. However, do not be deceived. This is merely the quiet before the storm. Genes do matter. They do not define our destiny, but they carry enormous information about our predispositions, our vulnerabilities, our disease risks, our personalities, our potentials, and our temperaments. A good rule of thumb is that genetics can explain about a quarter to three-quarters of the variation among humans in just about any trait we think of as important – whether we are shy, prone to cancer, tall, nurturing or quick-witted. For people living in typical environments in the developed world – with adequate nutrition and rearing – the environmental influences shared by the members of a family explain little of the variation of traits among us. Our differences depend more on the unique, individual influences that we cannot readily distinguish from mere chance. Genes are the biggest single window into who we are, and medical science is drawing back the curtains.

The biotech revolution encompasses far more than the elucidation of our genes. Systems-wide approaches to biology examine diverse groups of molecules that make up the complicated biological networks that constitute us. There are extensive efforts on the proteome, which is the complement of all our proteins; the glycome, which refers to the complement of carbohydrates; the metabalome,[3] which refers to the low-molecular-weight metabolites in cells; and even the kinome, which is the universe of protein-phosphorylating enzymes. Each of these is a nexus of intellectual activity to aggregate information and sift and analyze it for patterns that correlate with disease and health. This great ferment is the crux of our untangling of human biology and little of it could be occurring without the information technology (IT) that the silicon revolution has provided medicine to capture and analyze such data.

The references to the Human Genome Project in this chapter may be seen as encompassing the entirety of the attendant ferment in molecular biology. The elucidations of genetics will be felt first, because it is far simpler than realms like the proteome. The path from genotype to adult phenotype may be hard to lay bare, but at least the genome is a linear structure, and changes to it can be tabulated, analyzed and compared with resultant changes in adult phenotype.

The wave of changes to medicine from the Human Genome Project will arrive in ten to fifteen years. The biggest challenges will come not from specific tests for particular genes correlated with particular diseases, but from testing and analyzing our entire genomes. This will depend both on the arrival of cheap, reliable whole genome testing and on figuring out what the information actually means. A personal genetic profile is worthless unless combined with knowledge that is virtually non-existent today. However, once the technology for cheap, comprehensive testing exists, large populations will be tested and such knowledge will begin to emerge. These new understandings will expand the role of IT in medicine, shift the balance of knowledge between physician and patient, and change our perceptions of health and disability.

One of the most significant changes ahead will be a massive shift towards preventive medicine. As we begin to understand what various constellations of genes tell us about our disease vulnerabilities and predispositions, we will not ignore the information. Abstract general warnings are less gripping then concrete personal ones, and many people will want to ameliorate their risks through nutritional, lifestyle and pharmaceutical interventions. Yet, given how hard it is even to follow a healthy diet and get enough exercise, many people will probably want a pill.

The prospect of new preventives to help us maintain good health does not initially sound like it would shake the foundations of medicine. Look at how many people already take statins for heart disease or vitamins and nutritional supplements for that matter. However, the real challenge will not be the interventions, but the process of deciding what to do and how to communicate and validate this highly individualized information.

Most physicians will not understand the significance of such information about a particular patient. They are too busy for forays into so difficult and unfamiliar a realm of medical science. Consider cystic fibrosis, which can result from nearly 1,000 distinct mutations. One of these mutations is responsible for some 70 percent of cases, but at least 20 other mutations are each responsible for at least 0.1 percent of sufferers. Some of these mutations have different prognoses.[4] Wayne Grody, the head of the UCLA's DNA Diagnostics Lab, routinely spends hours explaining the implications of these tests to patients. He does so, not because it would normally be part of his job, but because the patients' primary physicians do not understand these details, as they see too few cases and have too many other demands on their time.

The most likely way to communicate complex information of this sort to large numbers of people will probably be a collaborative effort combining the Internet and specially trained genetic counselors. Yet, how will such a system be paid for? Where will the counselors come from? Where will the infrastructure be located? There are only a few thousand in the entire United States today. The transition to such a regime will greatly stress the existing patient–physician relationship.

We are at the leading edge of this today. Many doctors already encounter patients who lack a basic understanding of medicine but are nonetheless far better informed than they themselves about recent developments in the diagnosis and treatment of particular disorders. Small wonder, such patients have the ultimate motivation – their own survival. Selecting a treatment is thus becoming more a dialogue between patient and physician than a rendering of medical opinion from on high. Couple this change both with people's growing acceptance of "alternative" medicines and with the pharmaceutical industry's increasing use of direct-to-patient marketing, and the medical landscape will be very different from what it was even a decade ago.

Moreover, as more and more interventions become preventive and elective, huge political issues arise. Who will pay? Who will decide what is necessary? Who will have access? Such questions are particularly challenging in the United States, where private health insurance and health plans have short time-horizons ill-suited for long-term preventive care, but these developments are challenging as well where universal medical access predominates, because medical care still costs money. Populations are aging, taxes are high and the more useful such interventions turn out to be, the more people's expectations about their health will increase and the more they will want to spend. Our appetites for better health and the interventions to maintain it are bottomless. Our resources are not. To the extent that the costs of preventive interventions are recouped by reducing or delaying high-tech interventions, the underlying conflict is minimal, but where such savings do not occur, the choices are harder, because tradeoffs must be made. Increasingly, the lines between medicine, healthcare, and even social security policy will blur.

The advent of pharmacogenomics – the tailoring of drugs to our individual genetic backgrounds – is frequently touted as a development that will have large impacts on medicine. While this may bring more and better drugs, it will not greatly challenge the medical paradigms of today. Undergoing additional tests to guide our choices among alternative treatments is a qualitative rather than a quantitative change. The results of pharmacogenomics will be far more consequential in the pharmaceutical industry, where they may allow the revival of good drugs that have failed because of adverse side effects in some patients. If it becomes possible to identify and exclude these individuals beforehand, then the drug becomes viable. The same is true if a subgroup of particularly strong responders can be identified prospectively and singled out for a drug. As we move towards personalized medicine, however, it will become ever more challenging to do the testing that today ensures safety and efficacy. Thus, pharmacogenomics will stress the existing regulatory process.

The biggest impact of pharmacogenomics on medicine may be to speed the infusion of technology into decision making by adding a new layer of diagnostic tests and bringing a mass of new information into the medical record. This will necessarily increase reliance on expert computer systems to help formulate and verify prescriptions, a trend that is already well advanced.

The increasing ability of physicians to fine-tune pharmaceutical regimes will have another effect – it will draw us to turn to medication more quickly. Today, pharmaceuticals account for only 8 percent of total medical costs in the United States. Given how much cheaper drugs are than other interventions, cost pressures will push us towards greater reliance on pharmaceuticals. If drugs can keep an Alzheimer's patient out of a nursing home for longer, or shorten a hospital stay, the healthcare savings will be enormous.

This highlights one of the most difficult issues in medicine: the balance between individual privacy and collective progress. As animal studies and large clinical trials grow more difficult and more expensive, it will make even more sense to share medical records to find subtle drug interactions and develop better outcome predictions for patients clustered by genetic and symptomatic criteria.

The more we are able to capture and analyze huge amounts of information about our healthcare and medical histories, the more we will want to do so. After all, humanity is engaged in an immense, ongoing, collective medical experiment involving pharmaceutical, nutritional, lifestyle and other interventions affecting our health. At present, we throw away most of the data in that experiment because the cost of acquiring it, sifting through it, and analyzing it is prohibitive. If we have a specific question to ask, it would be better to design a clinical trial to find the answer. However, as the cost of acquiring and analyzing such data drops, and the ability to de-identify the information to hide its connection with any particular individual becomes widespread, while the cost of putting together a clinical trial increases, the economics will shift.

It is becoming more and more attractive to develop the institutional structures, record templates and networking tools to gather this information and sift through it for clues that will help us tease apart the many factors that interact to affect human health. The broader the information-gathering net is cast, the more interesting the results will be, because lifestyles, genetics, environment, nutrition and disease incidences vary so widely throughout the world. As individuals, we will want this information because it will mean so much to our health and will help us make better decisions about the risks and benefits of our various choices. As a society, we will need this information to better evaluate the larger choices we face about tradeoffs between injuries that arise directly when incompletely tested interventions go awry and those that arise indirectly when beneficial technologies are delayed.

It is hard to imagine that gathering and handling personal health data will not play an ever-larger role both in medical research and in the practice of medicine. However, this is a sensitive realm where matters of individual privacy and collective benefit may collide. There will be many political conflicts about how to proceed. The finalization of the landmark Health Insurance Portability and Accountability Act (HIPAA) in the United States in 2001 illustrates the layers of regulation that privacy worries can inspire and how such regulation can impact the practice of medicine by influencing healthcare infrastructures. This attempt to ensure the privacy of medical records by tightly controlling how such information can be used and shared was driven by many developments. Some of the most prominent were the digitalization of medical records and our growing capacity to aggregate large medical databases and access health records for purposes ranging from research and education to consumer marketing and patient solicitations for clinical trials.

There has been an attempt to downplay the political consequences of having broad easy access to our genetics. Scientists like to emphasize the near genetic equivalence of all humans, citing that 99.9 percent of the letters in our genetic codes are identical, even among those with very different origins and ethnicities. The figure is correct, but the message that we are all the same misses the point. In a genome of three billion base pairs, one difference in every thousand bases still totals three million differences. Furthermore, only a few such differences, if they are critical, can mean death.

Of course, we are all the same. We are all animals, all mammals, all primates and all humans. However, though we may all look similar to a hungry lion or an extraterrestrial visitor, our individual appearance differs. Human perceptions and sensitivities are finely tuned to see not the similarities but the subtlest of differences. This makes perfect sense, since we compete with each other for everything from money and power to friendships and sexual partners.

We will have to come to grips with the politics of this as we increasingly explore the actual biological differences among us, both as individuals and as human sub-populations. Heated controversies will undoubtedly arise. And these will have impacts on the practice of medicine, especially as we try to design pharmacogenomic interventions and organize studies to help tailor medical interventions for particular individuals, genders and populations. To deny the existence of such differences is not a good start.

Long-Term Developments in Medicine

The above short-term developments in medicine and healthcare are significant, but they pale alongside the longer-term changes that the revolution in molecular biology will bring. What if we learn to rework central aspects of our biology, unraveling the process of aging and understanding its genetics and biochemistry well enough to retard and control the process? What if we learn to fashion cocktails of drugs to manage our emotions in a targeted, nuanced fashion without the physiological side effects of today? What if we learn to control the genetics of an embryo so as to shape the temperament and personality of future children?

These are not wild fantasies. Mainstream researchers in the biology of aging believe that a doubling of human lifespan is plausible. Steven Austad, a leading gerontologist at the University of Idaho, is convinced that the first human who will reach the age of 150 has already been born.[5] Such a profound shift in the trajectory of human life would transform our relationships, our institutions, our vision of ourselves, perhaps even our concepts of meaning and purpose.

Medicine would necessarily be at the heart of these monumental changes. However, because the entire project of human enhancement does not fit easily into traditional medical models, it would be an uneasy adjustment. The idea of such enhancement does not sit well with many physicians, some of whom would oppose such possibilities whether or not they were safe and reliable. They see their role as healing the sick, and cannot imagine seeing aging and death as tantamount to disease, despite its being a debility that afflicts us all and that like any other "disease" raises morbidity. I suspect that were it possible to control the process of aging, however, attitudes would shift rapidly, and the underlying goals of medicine would have to adjust. These are the real changes that the biotechnology revolution may bring to the practice of medicine. We have our heads in the sand if we imagine that today's deciphering of the workings of life will merely bring us new treatments and preventives. It will call into question the whole nature of the project of healthcare by erasing the line between therapy and enhancement, though some bioethicists still cling to the above line and maintain that therapy is a laudable goal, while the struggle for enhanced function or added longevity is not.

Where does one draw that line? If the underlying process of aging could be slowed so that health could be prolonged for a few decades, this undeniably would be enhancement, but given that it also would hold off the diseases of aging that ultimately afflict us, it would be preventive therapy too. However, such disputation is academic, because many people would fight for it, if a drug existed to prolong youth. Without question, we would use such an intervention, and judging by our use of vitamins and cosmetic surgery, many people would like to be a little healthier, faster, stronger, smarter, more talented, youthful, and attractive. Indeed, such possibilities will wrench medicine.

A second challenging realm will be the development of drugs that shape and channel our moods and emotions. We are not strangers to such pills: the failure of the international War on Drugs is clear testimony to the attraction that these possibilities hold for us. Ritalin, which is used to treat ADHD (attention deficit/hyperactivity disorder), and the anti-depressant Prozac are but clumsy baby steps. Imagine what would happen if we could one day take a pill that makes us feel contented and fulfilled without the side effects that attend today's illicit drugs. Would we be able to resist such a pill? And if we could not, who would we be, and what would motivate us? Who would the "we" be that is deciding how to manage these various drivers of our behavior? This whole realm is termed cosmetic psychopharmacology, and it is bound to be one of the major medical challenges that emerge from the biobehavioral breakthroughs in the decades ahead. The crux of the matter is that our emotions have evolved to channel our behavior in ways that contribute to our survival. Love, anger, greed and envy motivate us, and if we short-circuit them and shape these feelings according to our whims, who would we be?

Reproductive medicine is a third realm where technology will bring profound changes to medicine. As we uncover the constellations of genes that help determine human potentials, vulnerabilities and temperaments, people will want to make choices about the genetic constitutions of their children. However, choosing our childrens' genes does not fit into traditional medical models. Even the idea is controversial.

There are three potential ways of effecting such choices. The first is reproductive cloning, which elicits enormous public interest but is little more than a media sideshow at present. The procedure has not been done in any primate, but is unlikely to remain confined for long to livestock, mice and the occasional pet, despite the many uncertainties and dangers that accompany it. It would be surprising if a human has not been cloned within a decade. The procedure will almost certainly happen before any mainstream researcher considers it safe, and a human cloning will almost certainly be announced many times before it actually occurs.

All the attention about cloning seems disproportionate to the threat. After all, when everything is said and done, a clone is but the creation of a delayed identical twin, and though such an occurrence may be strange and discomfiting, it will not topple civilization any more than "test-tube" babies did. Most of us have met identical twins and know that they are similar, yet different. Without question, they are unique individuals. Media interest is not entirely misplaced, however, because cloning has become a symbol for all the unfamiliar and troubling possibilities associated with advances in biotechnology.

Human cloning, which is yet to occur, has already exerted large effects on medicine. An entire realm of medical research, the use of nuclear-transfer procedures to create embryonic stem cells that might serve as transplant tissue, has been greatly hampered. Indeed, in 2002, the US Congress nearly passed a law to criminalize this so-called therapeutic cloning and punish transgressing medical researchers with ten years in prison.[6] That the proponents were untroubled by the idea of rendering basic biomedical research on cells a felony or by the idea of real people with real diseases and real suffering – people with Parkinson's, diabetes or spinal injuries – being denied the hope of this line of research, shows the strength of some people's philosophical and religious objections to cloning.

Ironically, while this debate rages, a far more potent way for parents to choose a child's genes – pre-implantation genetic diagnosis (PGD) or embryo screening – has been practiced for a decade. PGD is simple in concept. Medical technicians tease a cell away from a six- or eight-cell embryo without injuring the embryo, and test the cell's genetics. The prospective parents then use the test results to decide whether to implant the embryo or discard it in favor of another. Since the early 1990s, so many couples at high risk of having a child with cystic fibrosis or other serious genetic diseases have used PGD to avoid such infirmities that the procedure is not even controversial any more. However, it soon will be once again, because the technology is poised to allow couples to make choices about attributes entirely unrelated to disease.

Consider what prospective parents would be able to do if they could screen a hundred of their embryos for a predisposition towards some particular quality that intrigued them. The parents might be seeking a child who was outgoing, tall or even devoutly religious (the tendency to view religion as meaningful in one's life seems to be moderately heritable). Or perhaps a couple might want to have a child with a high IQ, a measure of intelligence and aptitude, which is more than 50 percent heritable.[7] If a couple in which both members have an average IQ could identify and implant the one embryo out of a hundred most likely to manifest a high IQ, the couple would on average bear a child with an IQ twenty points higher than the average score of one hundred within the general population. Such children would score higher than 90 percent of the population, and this would occur through a safe, reliable procedure that might be feasible in a couple of decades. It does not sound like medicine, but it is likely to become so, just as in vitro fertilization has moved from the fringes of science to the mainstream of reproductive medicine. Clearly, the potential challenges of PGD are extraordinary.

Anyone familiar with in vitro fertilization (IVF) – the laboratory conception of embryos – is no doubt aware that couples undergoing it cannot produce a hundred embryos. A woman is lucky to get even a dozen good eggs after repeated injections of hormones in a difficult and unpleasant process that few couples would undertake except to overcome infertility. Yet, this is IVF today. Tomorrow, IVF may be much easier. Once it becomes possible to harvest immature eggs, freeze them, and later thaw and mature them in vitro, the picture will change. A woman could harvest thousands of eggs using a simple ovarian biopsy, and freeze and bank them. Years later, she could thaw the eggs, mature them in vitro, fertilize them with her partner's sperm, and implant one that has either been selected at random or screened for diseases and other traits.

Once this is feasible, such screenings would be used first by the infertile, then the wealthy and then everyone else. Of course, people might screen initially only for serious diseases, but soon they would probably be looking for lesser vulnerabilities like a tendency towards manic depression, and then a plethora of other attributes. In an international poll[8] compiled in the early 1990s by Darryl Macer, a bioethicist at the University of Tsukuba in Japan, anywhere from 24 percent (in Japan) to 80 percent (in Thailand) of people from around the world said they would enhance the physical or mental capabilities of their future children if it were possible.

A third method of choosing our children's genes – directly altering those in the first cell of the embryo – makes simple embryo screening seem tame. Such so-called "germline" engineering, which refers to the reproductive or germinal cells, is the most direct way of manipulating human genetics. Its advent will signal the beginning of conscious human design, and it may be only a generation or two away.[9]

One reason why direct germline manipulation is so likely to arrive is that its early applications will be indistinguishable from those of embryo screening. Does it really matter if prospective parents screen several of their embryos to find one free of a fatal cystic fibrosis mutation or alter an embryo to correct the mutation? Does it really matter if a couple selects an embryo predisposed for high IQ or alters one to create the same predisposition? Deciding whether to screen or engineer embryos may depend, not on ethics, but on engineering issues like safety, practicability and reliability. Given the painful nature of mistakes for everyone involved, most parents will probably be very cautious about "enhancement" technologies.

It is critical to understand that such technologies will not be creating superhumans any time soon. Within the typical range of expression of any trait, improving the likely potential of a future individual who would otherwise be average or below average will be far easier than improving an individual whose potential would already be high. Thus early interventions will help those with the least potential, and the net effect of such enhancement will be to raise the performance of those who would otherwise be at the bottom end of the distribution. Early enhancement will be more egalitarian than elitist.

We deceive ourselves if we think that future parents will not be facing wrenching, difficult choices about the genetics of their children. And they will face these dilemmas whether they decide to use such technology or to forgo it, because they will know that their children will judge their decisions. Moreover, medical science will be at the heart of the controversies that swirl around these coming challenges in reproductive biology.

Meeting Future Challenges

As molecular biotechnology begins to offer humanity a smorgasbord of new possibilities such as life extension, cosmetic psychopharmacology and genetic engineering, it is unlikely that the medical establishment will embrace the pursuit of these transformative developments, but it will not be able to ignore them for long.

We must not be blinded, however, by the amazing new knowledge that technology is bringing medicine, for it is worth looking at another revolution underway today, one that is seemingly at odds with these high-technology possibilities. Throughout the developed world, there is renewed interest in traditional medicine – the medicinal herbs and remedies from the human past. This development, however, is less a counter-revolution opposed to modern medicine than one might imagine, because technology is driving this development as well. Our exploding capabilities in information sharing and processing are bringing together knowledge from all over the world about our interactions with the rich chemistry in the biota surrounding us. The value of this information cannot be overstated. Human biochemistry differs very little from that of any other animal, so it is hardly surprising that we can be profoundly influenced by this biota. After all, different forms of life have been trying to influence one other throughout evolutionary history.

The powerful new screens and tools of modern biochemistry are precisely what will allow us to extract valuable new drugs from the noisy signal of alternative medical concoctions. Though obscured by anecdotal information and uncontrolled observation, this biota so far has given us about half of our pharmaceuticals.

The integration of these two realms offers a potential solution to the challenges posed by the following:

- the arrival of widespread preventive medicine;
- regulatory bottlenecks that are retarding the arrival of new drugs;
- price pressures on healthcare.

Signum Biosciences, for example, is attempting just this integration by using advanced proprietary screens to extract botanical drugs from medicinal herbs. The initial focus is on Alzheimer's, diabetes, heart disease, and other illnesses where key regulatory problems substantially add to disease risk. The company's goal is to identify, test and characterize a group of effective, affordable botanical extracts that can supplement or even replace certain pharmaceutical interventions. And such preparations face greatly lowered regulatory hurdles, so they can reach patients much more rapidly.

As technology becomes increasingly powerful, medicine will always face the challenge of making such technology serve us and of integrating it with existing knowledge and practices. Not everything that can be done should or will be done, but with basic human aspirations such as having extra years of health, managing our emotions, or giving advantages to our children, the question is not if, but when, how and where these will happen.

Humanity will go down these high-tech paths for two reasons. First, because the technologies will be mere spin-offs of mainstream medical research that enjoys widespread and enthusiastic support. Second, because we are human. We tinker, explore and try to improve our lives. We plant, build, dam, hunt and mine, and do so on an increasingly large scale. For as long as it has been possible, we have also altered ourselves. We pierce our bodies, cut our hair, tattoo our skin, straighten our teeth and fix our noses. We use drugs to reduce pain, lose weight, go to sleep, stay awake or just get high. To imagine that we will forgo new and better ways of modifying ourselves is sheer denial of past indications. Holding back such technology for long would require global totalitarian repression worse than anything that the technology itself might bring.

Medical science – the gatekeeper for such technologies – will necessarily be jarred and buffeted by them. Yet, we might as well face it. Our next frontier is not space but ourselves, and medicine and biotech will be leading this journey of exploration.

Of course, some scientists say these things will never happen. They maintain that our biology and genetics are too complex to modify and adjust usefully. However, while some manipulations undoubtedly will be impossible, others will merely be difficult, and still others will be surprisingly straightforward. It is easy to find simple biological interventions that bring huge effects. Altering a single roundworm gene can double the worm's lifespan, which is something no one would have guessed a few decades ago. Fire ants provide an even more remarkable example. Changing a single gene involved with pheromone detection can switch a fire-ant colony from having one queen to having many – a complex social behavior undoubtedly linked to many genes and biochemical pathways.

Various critics claim that we can simply ban these technologies. We probably will, particularly now, while the technologies do not yet exist. Ultimately, this will merely shift development elsewhere, drive it from view, and reserve the technology for the wealthy, who can most easily circumvent such restrictions. Embryo screening is illegal in Germany, so affluent Germans simply seek it in Brussels or London.

The moral challenge for medicine today is whether we will face these emerging possibilities and develop sane, pragmatic oversight for them or push them off on our children. This is not a battle between right and wrong, but between right and right, because these new technologies will bring great benefits and great losses. Our biggest choice may be whether we see our choices as being made collectively and imposed on the population, or as being left to individual discretion. Different societies are bound to answer this question differently.

Some people believe that with enough education and discussion we will be able to reach consensus about how to handle these technologies. But this is extremely unlikely, no matter how long we talk. There will be no consensus, because these matters touch our values too deeply. Our views depend too much upon history, philosophy, culture, politics, religion and temperament. A difficult struggle is approaching and its two poles are clear. Some will see these technologies as the invasion of the inhuman, an abomination that will assault human dignity and everything humanity has struggled to build over the millennia. Others will see them as the crown of human achievement – our lucky chance to transcend biological limitations in ways that other generations could only dream of.

These technologies will challenge our vision of who we are and try our willingness to face an open-ended, uncertain future. Ultimately, our real test is not how we deal with cloning, genetic screenings, artificial hearts, GM foods, or any other specific technology, but whether we continue to embrace the possibilities of the future or pull back in fear and relinquish their development to other, braver – or simply less cautious – souls in more adventurous regions of the world. China and a number of other nations are not about to forgo these technologies.

What is the message of religion here? "Don't play God!" could not be it, because we already do so whenever we take antibiotics, get a transplant or board an airplane. The biggest irony is that this new path opening before us is a very spiritual one. Embarking on this voyage to we know not where is an extraordinary act of faith. It must be so. This self-directed evolution is not something we can plan and control. Its course hinges too much on the character of future technologies that we cannot yet glimpse, and the values of future humans that we cannot understand.

We have entered a new millennium, and long before its end, perhaps in only a century or two, future humans – whoever or whatever they are – may look back at this moment, and see it as a turbulent, difficult moment. They also will see it, though, as a unique, glorious instant, when the very foundation of their way of life was laid. One of the cornerstones will be the reworking of human biology. It will come gradually at first and pick up speed as the biotechnology revolution bears fruit. Ultimately, it will do more than remake medicine. It will challenge our very notions of what it means to be human.

In my view, it is a remarkable privilege to see this critical transition in the history of life. However, we are more than mere observers of this broad set of fundamental changes that will soon reshape the trajectory of human life and our understanding of it. We are also its architects and its objects. This is why it will be such a challenge.

In 430 BC, Thucydides eloquently pointed out the challenge that faces us, medicine and humanity as a whole in the next century: "The bravest are surely those who have the clearest vision of what is before them, glory and danger alike, and yet notwithstanding, go out to meet it."

11

Biotechnology and the Future of the Pharmaceutical Industry

Allan B. Haberman

The global pharmaceutical industry in 2003 finds itself in an era of tremendous opportunity and challenge. On the one hand, life scientists have given the industry a host of new technologies and scientific discoveries with the potential to be developed into breakthrough treatments and even cures for humanity's gravest diseases. On the other hand, the industry's traditional drug discovery methods have reached the limit of their ability to yield innovative new drugs. This is compounded by the growing number of large-selling branded drugs facing patent expirations during the 2000–10 decade, thus subjecting them to competition from generic drugs. The industry has therefore invested heavily in the new technologies, and has planned to use them to discover and develop new therapeutic agents to make up for its shortfall in high-valued drugs. However, pharmaceutical companies have not yet been able to use the new technologies effectively and consequently are left with a deficit of the new drugs that they need to fill their pipelines. The industry is also confronted with a set of societal pressures, especially the current demands to control drug prices and the demands of the stock market for ever-increasing shareholder value, especially as determined by increasing quarter-to-quarter earnings.

Pharmaceutical companies have attempted to meet the challenges of the current era via business strategies such as in-house investment in advanced drug discovery and development technologies, forming partnerships with biopharmaceutical and technology companies, corporate restructuring, and industry consolidation via mergers between large pharmaceutical companies.

Arguably the most salient feature of the pharmaceutical industry landscape in the late 1990s and early 2000s has been the waves of mergers between Big-Pharma companies that have continued to consolidate the industry. This has culminated in the US$60 billion (all-stock) acquisition of Pharmacia by Pfizer, which was initiated in 2002 and completed in April 2003. This mega-merger created a pharmaceutical giant with global prescription drug sales of over US$48 billion (approximately 11 percent of the worldwide pharmaceutical market), and United States sales of over US$25 billion. The Pfizer/Pharmacia deal follows a previous Pfizer mega-merger in 2000, in which the company acquired Warner-Lambert.

As a result of the merger with Warner-Lambert, Pfizer acquired a number of blockbuster drugs.[1] This includes the blockbuster drug Lipitor (atorvastatin), which is the world's largest-selling statin drug for the treatment of high blood cholesterol.[2] Lipitor had worldwide sales of approximately US$6.4 billion in 2001. In merging with Pharmacia, Pfizer acquired the blockbuster arthritis medication Celebrex (celecoxib), which had 2001 sales of over US$3.1 billion, as well as Bextra (valdecoxib), a newly launched second-generation COX-2 (cyclooxygenase) inhibitor, which is a drug of the same class as Celebrex and Merck's competing drug Vioxx (rofecoxib).

In addition, the merger resulted in Pfizer's acquisition of Pharmacia's congestive-heart-failure drug Inspra (eplerenone). Inspra is a potential blockbuster that was approved by the US Food and Drug Administration (FDA) for high blood pressure in September 2002. In April 2003, Pfizer submitted an application for the approval of Inspra for prevention of post-myocardial infarction heart failure. The merger makes Pfizer the number one pharmaceutical company (in terms of sales) in the world, and the leader in the cardiovascular and arthritis and pain markets. It also gives Pfizer a strong position in cancer and infectious disease.

Pfizer's merger strategy is motivated by the need for the company to sustain its high rate of growth, both in annual revenues and in earnings per share. According to company management, the merger will enable Pfizer's earnings per share to grow to 19 percent over the next three years, as compared to 16 percent without the merger.

Pfizer's growth has depended on cost savings from company mergers – that is, by eliminating redundant staff and functions in the two merging companies. Pfizer expected US$1.6 billion in cost savings over three years from the merger. However, the growth via cost savings from the merger is winding down, and Pfizer's growth has fallen from 18 percent in the first quarter of 2001 to 10 percent in the second quarter. As a result of the Pharmacia merger, Pfizer expects to realize US$2.5 billion in cost savings.

However, the real engine of growth, especially over the long term, is a company's portfolio of patent-protected marketed drugs, and its research and development (R&D) pipeline. Except for the blockbuster drug Viagra (sildenafil citrate) the largest-selling drug for erectile dysfunction, which Pfizer developed in its own laboratories, all of the company's largest-selling drugs – Lipitor, the epilepsy treatment Neurontin (gabapentin) and the allergy drug Zyrtec (cetirizine hydrochloride) – were either acquired via mergers or in-licensed from other firms. In 2002, Pfizer expected to submit just one or two drugs for approval by the FDA, despite a 2002 R&D budget of US$5.3 billion, which is expected to rise to US$7 billion after the merger with Pharmacia.

In 1991, the former head of Pfizer's R&D, John Niblack, said that Pfizer could only bring two drugs to market per year, which is not enough to sustain growth of revenues and earnings at a rate expected by investors. He also said that Pfizer is not the only Big Pharma facing this problem.[3] The acquisition of Pharmacia thus buys Pfizer breathing space, until it can find ways to improve the productivity of its R&D, or, alternatively, complete another merger that give the company ownership of large-selling drugs. However, at some point Pfizer will need to be able to develop most of its own drugs and in-license and form partnerships (for example, with biotechnology companies) for others, since the larger the company gets, the more drugs it needs to launch every year.[4] Pfizer also faces several patent expirations of its drugs during the period 2004–07 and will need to replace them with new drugs to maintain its growth.

Moreover, the steady expansion of Pfizer (and consequently its rising sales base) means that it must develop and market more and more high-valued drugs to sustain its growth rate, in percentage terms. This means that the company must focus its development efforts on drugs expected to generate a large volume of sales. After the merger with

Warner-Lambert, Pfizer's criterion for drug development targeted agents that were expected to generate at least US$500 million per year.

This is not out of line with other Big Pharmas, most of which also have large sales bases resulting from mega-mergers, or in some cases due to internal growth (notably Merck, the world's largest Big Pharma that has not participated in mergers). For a Big Pharma company to meet investor expectations and achieve annual revenue growth rates of 10 percent, it must on average launch four new molecular entities (NMEs) per year, each with average annual sales of US$350 million. NMEs are also known as "new chemical entities" (NCEs).[5] However, from 1996 to 2001, Big Pharmas launched on average less than one NME per year per company. Moreover, of the NMEs launched in 1996, only approximately 25 percent had achieved sales of US$350 million or higher by 2000.[6] These are some of the daunting odds that are motivating consolidation in the pharmaceutical industry.

Despite the sales goals of Big Pharma companies, many drugs for important human diseases, at the time of their development, are expected to generate significantly less than US$500 million per year or even US$350 million per year, even though some of these drugs may reach US$500 million in the post-marketing period. In other cases, a drug may be developed first for one indication, for which it is easiest to obtain approval by regulatory agencies. In this situation, the initial market for the drug may be small. Drug makers then conduct new clinical trials for the drug in other indications, attempting to expand the market for the drug. Of course, there is no guarantee that these additional clinical trials will be successful.

Drugs that are expected to address small markets, or which initially address small markets but might expand their markets after a long process of further development, are not good candidates for development by the largest pharmaceutical companies. However, for a smaller company, such as most biotechnology companies, developing a drug that brings in US$200 million per year represents a major success, and may well result, for example, in turning an unprofitable development-stage company into a profitable company, or turning a one-product company into a more stable company with multiple products and greater revenues. This is one reason why many new drugs are developed by smaller companies; other reasons for this trend are discussed in later sections.

The Pfizer–Pharmacia merger has created a corporate giant whose magnitude threatens to overwhelm other Big Pharmas, especially in terms of the size of the expanded Pfizer's R&D budget and sales force as compared to its competitors. Many industry commentators therefore expected the announcement of the Pfizer–Pharmacia merger to trigger a wave of additional mega-mergers. However, this has not yet happened, and others are not as certain that it will happen in the next several years. Pfizer contracted the merger with Pharmacia from a position of strength, not only in terms of revenue growth, but also because, unlike many of its rivals, most of its major products have not recently been affected by patent expirations. Moreover, Pharmacia is also a strong company, in terms of its product portfolio and its patent protection.

However, several other Big Pharmas that might be contemplating mergers are doing so to deal with problems, such as recent patent expirations, recent failures or delays in the approval of key late-stage drug candidates, and manufacturing and regulatory difficulties. For example, Eli Lilly has been affected by the expiration of the patent for its blockbuster antidepressant Prozac (fluoxetine), and FDA approvals of several of Lilly's drugs were delayed due to manufacturing quality-control problems.[7] Schering-Plough has also been hit with the expiration of the patent for its blockbuster antihistamine Claritin (loratadine), and it also lost a court case in August 2002, which allowed generic versions of loratadine to reach the market, beginning in December 2002.[8] Schering-Plough has also had manufacturing compliance problems with the FDA. Bristol-Myers Squibb (BMS) is threatened by several patent expirations, has not been able to get a drug approved by the FDA in two years, and in-licensed its largest-selling drugs, rather than developing them in-house; thus its internal R&D has not been very productive. Moreover, BMS entered into a US$2 billion partnership with the biotechnology company ImClone, to develop and commercialize ImClone's promising cancer drug Erbitux. However, in a widely publicized case, the FDA refused to consider the application for approval of the drug, because of poorly designed clinical trials.[9] These and similar problems make these companies poor candidates for mergers that can be expected to have good business outcomes.

From an industry-wide standpoint, the potential for more mergers is limited by the reduction in the number of Big Pharma companies, which is the result of mergers that have already occurred. Moreover, the real targets of mergers are blockbuster drugs with several years remaining for their patent expiration. Since the number of these drugs, both currently marketed drugs and late-stage potential blockbusters, is relatively small, this dampens the appetite of Big Pharma companies for mergers. Nevertheless, several large pharmaceutical companies are reported to be contemplating mergers, so a few additional mergers between Big Pharmas during the next two years are not completely ruled out.

When many pharmaceutical executives, financial analysts and other industry commentators speak of the "future of the pharmaceutical industry," they are often referring to predictions of future industry consolidation via mergers, and/or the rise of new competitors to the Big Pharmas. This could happen, for example, via biotechnology companies growing to Big Pharma status by a combination of internal development, mergers and acquisitions, or by becoming niche players that develop portfolios of drugs that address important diseases, but that have small markets. However, the main factor that determines pharmaceutical industry structure is the difficulties that companies face in developing sufficient drugs, especially high-valued drugs, to support their growth. We shall therefore examine what causes these difficulties in drug development, what companies might do to overcome such difficulties, and the enabling role that biotechnologies can play in this regard.

Improving the Time and Efficiency of Drug Development

There are two key related factors that affect the ability of pharmaceutical companies to develop new drugs: the time and cost to bring a new drug to market, and difficulties in discovering high-quality drug candidates to enter a company's pipeline. According to the Tufts Center for the Study of Drug Development, as of 2001 it took about 10–15 years on average to bring a new prescription drug to market, at an average cost of US$802 million in 2000.[10] These average costs include R&D expenses resulting from drugs that fail to make it through the drug development process. The cost of developing new drugs is skyrocketing at a rate much greater than the rate of inflation. For example, if costs had increased only as the result of inflation, the cost to develop a new drug would have risen from US$231 million in 1987 to US$318 million in 2000.

Drug development occurs in three steps: drug discovery, drug development (which mainly involves clinical trials) and the approval process. In the United States, entry of a drug into clinical trials requires approval of an Investigational New Drug Application (IND). Upon completion of clinical trials, a company files a New Drug Application (NDA); the approval process begins when the FDA accepts the NDA for review. The major cost of developing a new drug is the cost of clinical trials; it is the rise in costs of clinical trials that is the major driver in the rising cost of bringing a new drug to market.

According to the Tufts Center, the time for FDA review and approval of new drugs has shortened substantially in recent years. The average approval time for priority-reviewed NCEs and new biopharmaceuticals decreased from 11.9 and 12.6 months, respectively, in fiscal years 1994–97 to 7.1 and 8.0 months, respectively, in fiscal years 1998–2000.[11] This accelerating of the approval process was made possible by the Prescription Drug User Fee Act, which was passed by Congress in 2002 and renewed in the 1997 FDA Modernization Act (FDAMA), with the provision that the FDA improve its responsiveness and communication with pharmaceutical companies. The user fees were designed to finance the expedited review process. Although review times for priority-reviewed NCEs and biologicals fell substantially between fiscal years 1994–97 and 1998–2000, the results for standard-reviewed NCEs and biopharmaceuticals were mixed. Review times for NCEs fell 9 percent between the two periods, but review times for biopharmaceuticals rose by 33 percent. Priority-reviewed NDAs are those that the FDA believes will have the greatest potential therapeutic value, and are thus slated for expedited review. Standard-reviewed NDAs are deemed to have little or no therapeutic advantages over existing drugs.

In addition to the shortening of the average review times for new drugs, the average time for a drug to go through clinical trials is shortening, as the industry finds ways to make the process more efficient. According to the Tufts Center, the mean clinical phase of development (i.e., the time from the filing of an IND to the submission of an NDA) was 19 percent faster in 1993–98 as compared to 1993–95.[12] Industry efforts to improve clinical trial efficiency include identification and redesign of poorly designed clinical trials, greater use of FDA advice on clinical trial design early in clinical development, process re-engineering, and improved data management. The FDAMA also allows for accelerated "fast track" approvals of drugs for certain serious or life-threatening diseases, which have the potential to address unmet medical needs for this condition. Under "fast track" procedures, the FDA may approve a drug for which good surrogate endpoints have been determined by well-controlled clinical trials. Such surrogate endpoints may include, for example, tumor shrinkage or elimination in the case of a cancer drug, or improvements in CD4 cell counts in the case of an AIDS drug. The manufacturer will then be required to conduct post-marketing Phase IV clinical trials after initial approval and marketing of the drug, in order to determine the effects of the drug on prolonging patient survival. This fast-track system has the effect of shortening the development time of drugs that address unmet needs in the case of serious or life-threatening diseases.

Despite the above types of improvements in the efficiency of drug development, the major reason for the gap in the ability of pharmaceutical companies to develop drugs lies in the area of drug discovery. More effective drug discovery may also result in a decrease in expensive late-stage failures in clinical trials, by weeding out compounds that are likely to fail early in the drug development process, and by focusing companies on biological mechanisms that are likely to lead to greater efficacy and specificity in treatment of disease.

It is drug discovery that has been impacted the most by the revolutionary changes in biotechnologies. It is also the area that will be the most impacted by improvements in the ability of companies to utilize the newer biotechnologies such as genomics and proteomics more effectively, resulting in greater numbers of viable drug candidates that address unmet medical needs. We shall therefore examine the limitations of the pharmaceutical industry's traditional drug discovery strategies, the nature of the new biotechnologies, and the impacts of these new technologies on drug discovery and on the pharmaceutical industry in general.

The Diminishing Returns of Traditional Drug Discovery Strategies

Why is the pharmaceutical industry unable to develop sufficient numbers of high-valued new drugs to meet its growth needs? A key factor is that the industry's traditional small molecule drug discovery strategies and technologies have reached the limit of their ability to yield new NCEs. Recombinant DNA and monoclonal antibody (MAb) technologies (which are discussed in the next section) have resulted in the launch of a growing number of protein and other large molecule drugs every year. However, these are often developed by biotechnology companies rather than Big Pharma companies. In some cases, Big Pharmas partner with biotech companies in order to co-develop or market large-molecule drugs, or Big Pharmas may acquire a controlling interest in a biotech company that produces large-molecule drugs,[13] or acquire the company outright.[14] Even so, revenues to Big Pharmas from large-molecule drugs do not fully make up for the gap in NCEs. Moreover, until improved drug delivery methods are developed for large-molecule drugs such as delivery via inhalation or minimally invasive implanted devices, and eventually, in some cases, by gene therapy, orally available small-molecule drugs are preferable for most diseases. This is because of their convenience, ease and safety of administration, and lower cost.

Despite the large investments in genomics and other advanced drug discovery technologies by the pharmaceutical industry, the majority of drugs introduced to the market in recent years were still developed by traditional drug discovery strategies, in some cases combined with the applications of the results of novel biological and biochemical research.

Traditional drug discovery is largely based on the modification of older drugs and drug leads to obtain new drug candidates, and a good deal of serendipity. This is augmented by the results of biological and biochemical research, such as studies of the biochemistry and biology of receptors and enzymes. Currently marketed drugs (which, except for a relatively few biotech-derived drugs, were all discovered via the traditional drug discovery process) target approximately five hundred molecules. The vast majority of these targets fall into a few protein families: G-protein coupled receptors (i.e. GTP-binding protein coupled receptors), ion channels, certain types of enzymes (e.g. serine proteases), and nuclear hormone receptors (e.g. steroid receptors).[15] Of these target families, G-protein coupled receptors are by far the most frequent targets of existing drugs – drugs that target these receptors account for 60 percent of pharmaceuticals on the market, with annual worldwide sales of over US$85 billion.

Utilizing mainly traditional drug discovery methods, even augmented by the use of high-throughput screening (HTS, an automated method, described in the Appendix, of screening very large numbers of chemical compounds to find potential drug candidates), the pharmaceutical industry has not been able to increase its output of NCEs over the course of the 1990s.[16] As discussed earlier, the present output of NCEs is insufficient for the growth of pharmaceutical companies, especially when they are affected by patent expirations or late-stage failures of drugs in development. This situation can lead to gaps in their pipelines, with adverse business consequences.

The fact that the traditional methods of drug discovery target so few families of molecular targets suggests that these methods may be "boxed in" in terms of the types of targets that they can address with new drugs. Drug discovery researchers therefore are looking to the new biotechnologies to enable them to address the much larger number of drug targets (approximately five to ten thousand) estimated to exist in the human genome.

The Continuing Revolution in New Technologies for Drug Discovery

Recombinant DNA and Monoclonal Antibody Technology

The development of modern biotechnology began in the mid-1970s, with the development of recombinant DNA technology and monoclonal antibody (MAb) technology. As the result of continual improvements in these technologies, and the development of a new class of "biotechnology companies" beginning around 1980, a whole suite of new biotechnology-based products has emerged on the market. These are mainly protein drugs that are either recombinant versions of natural proteins that replace older products made from materials in short supply or that have a high likelihood of viral contamination, or are completely new therapeutic proteins that were previously unavailable. Examples of recombinant protein drugs developed to replace older non-recombinant protein drugs include recombinant human insulin, human growth hormone and factor VIII.[17] New therapeutic proteins developed via recombinant DNA technology include recombinant erythropoietin and interferon products for the treatment of multiple sclerosis.[18]

Although the development and marketing of these drugs began in the early 1980s, the introduction of protein drugs based on recombinant DNA technology continues in the present, and is expected to be an important factor in the future. This is especially true of the accelerating number of MAb drugs being introduced into the market. The application of advanced recombinant DNA technologies to the production of MAbs made possible the current generation of successful MAb drugs; these have sequences that are largely or completely human, as opposed to earlier, unsuccessful versions that were of mouse origin.

Table 11.1 lists selected important biotechnology drugs that have been launched onto the market in the past five years. All but one of these (Novartis' small-molecule drug Glivec/Gleevec, which is discussed in a later section), are protein drugs derived from recombinant DNA technology or a combination of MAb and recombinant DNA technology. Many more MAb drugs are now in the pipeline, including late-stage drugs that may reach the market in 2003. Examples of recently introduced recombinant protein therapeutics include Amgen's Aranesp (darbepoetin alfa) a second-generation recombinant erythropoietin that can be administered less frequently than Amgen's Epogen, and Biogen's Amevive (alefacept, approved by the FDA in January 2003), which is the first biological treatment for psoriasis to reach the market.

Table 11.1
A Selection of Important Biotechnology Drugs Launched in the Last Five Years

Drug	*Company*	*Indication*	*Type of Molecule*
Amevive	Biogen*	Psoriasis	Recombinant protein (fusion with immunoglobulin G1)
Enbrel	Amgen/Wyeth	Rheumatoid arthritis	Recombinant protein (fusion with immunoglobulin G1)
Glivec/Gleevec	Novartis	Chronic myelogenous leukemia; gastric cancer	Small molecule
Herceptin	Genentech/Roche	Breast cancer	MAb
Natrecor	Scios/Johnson & Johnson	Congestive heart failure	Recombinant protein
Remicade	Centocor/Johnson & Johnson	Rheumatoid arthritis; Crohn's disease	MAb
Rituxan	Idec/Genentech	Non-Hodgkin's lymphoma	MAb

Note: *See Note 20 on Biogen's merger with Idec.

Source: Haberman Associates.

Except for Novartis' Glivec/Gleevec, all of the drugs listed in Table 11.1 were developed by biotechnology companies. However, Roche, as a result of its controlling interest in Genentech, markets the latter's Herceptin outside of the United States. Centocor, following its acquisition is a subsidiary of Johnson & Johnson. Wyeth co-markets Enbrel with Amgen, which acquired the drug via the acquisition of its developer, Immunex. Also, as the result of its April 2003 acquisition of Scios for US$2.4 billion, Johnson & Johnson will market Scios' Natrecor (nesiritide), a recombinant protein drug for acute congestive heart failure, which was launched in 2001. Johnson & Johnson is operating Scios as a subsidiary, as is the case with Centocor. In these various ways, Big Pharmas realize revenues from several of the drugs listed in Table 11.1.

Genomics, Proteomics, and Other Advanced Drug Discovery Technologies

In the 1990s and early 2000s, a new suite of complementary technologies emerged, which continues to revolutionize the practice of drug discovery. These include genomics, proteomics, functional genomics (and functional proteomics), combinatorial chemistry, high-throughput screening, structure-based drug design, bioinformatics, and related technologies.[19] For all these technologies, automation, miniaturization and parallelization are important in increasing the speed and efficiency and lowering the unit cost of such processes as genome sequencing, compound synthesis and screening, and assays.

The concept of "high-throughput" operations (that is, the ability to run large numbers of assays per unit time, for example 100,000 assays per day in ultra-high-throughput drug screening) is important in the use of advanced drug discovery technologies. This has led several companies and industry commentators to speak of "the industrialization of drug discovery." The technologies in this group are usually used together, as part of industrialized processes, to increase the efficiency of genomics-based drug discovery. The principal advanced drug discovery technologies are described briefly in the Appendix.

Effect of New Technologies on Industry Structure

The application of recombinant DNA technology not only resulted in new classes of drugs but also in the rise of a new set of competitors for the large pharmaceutical companies. Several of these have become large and successful, and are often referred to as "Big Biotech" companies. These include Amgen, Genentech (now majority-owned by Roche, but run as an independent, publicly traded company), Chiron, Biogen,[20] and Genzyme. All of these companies started as small, venture-capital-backed start-ups in the late 1970s or early 1980s, and grew into large profitable publicly traded companies with marketed products, including several large-selling ones. Other recombinant-DNA-based companies started at the same time were not so successful, and either were acquired, went out of business, or continued as small companies and attempted to become successful by changing their business models. One such company, originally known as California Biotechnology, was transformed into Scios, and changed its business model to one more in tune with the late 1990s and early 2000s. As discussed earlier, Scios (a Johnson & Johnson subsidiary since April 2003) launched a successful recombinant DNA-based product, Natrecor in 2001. Most of its ongoing research efforts are based not directly on recombinant DNA technology, but on genomics, combinatorial chemistry and studies of signal transduction pathways. "Signal transduction" refers to complex biochemical pathways that control responses of cells to their environment, as well as other aspects of cellular physiology.

The large biotechnology companies have not been without their rounds of consolidation. Amgen acquired Immunex, another Big Biotech company, in 2002. During the 1990s, Genentech was acquired by Roche, but kept as a majority-owned, publicly traded company with its own independent management. Also in the 1990s, American Home Products, now known as Wyeth, acquired Genetics Institute.

The early recombinant DNA-based biotechnology companies, launched in the late 1970s and early 1980s, provided the model for the "biotech company," which applies to newer companies whose technology is based on genomics, proteomics, combinatorial chemistry, and various types of advanced biological and chemical science such as signal transduction pathways, tissue engineering, structure-based drug design, and many others. Many of these companies are developing small-molecule drugs, like the Big Pharma companies. However, they utilize their proprietary technology platforms (often spun off from academic laboratories) to develop their small-molecule therapeutics. Moreover, the recombinant DNA technology-based Big Biotech companies are also moving into small-molecule drug discovery as well.

Big Pharma companies are also attempting to utilize genomics and other advanced drug discovery technologies to develop small-molecule drugs – as discussed earlier, they are in fact depending on this strategy to build their pipelines. They access these technologies in two ways: making large investments to build in-house genomics, proteomics and high-throughput screening facilities, and by acquiring or partnering with small biotech companies. For reasons discussed in later sections, the biotechnology industry tends to be more productive in the discovery of targets and drugs than are Big Pharmas. However, Big Pharmas are stronger in clinical development both in terms of having the large amounts of cash needed for clinical trials, and in terms of expertise, and also in marketing. This has led many industry commentators to call biotech companies the discovery engine of the pharmaceutical industry.

Big Pharmas may partner with biotech companies at any stage of the drug development process. Biotech companies can thus not only be competitors to Big Pharmas, but also collaborators that help the larger companies fill their pipelines. For example, these collaboration agreements may involve the following:

- acquiring access to validated targets, against which the Big Pharmas screen their large libraries of compounds;
- co-development of drugs originally discovered by a biotech company in clinical trials;
- co-marketing deals, in which a small biotech company takes advantage of the marketing strength of a Big Pharma to increase the sales of a drug that it has developed.

Biotech companies that partner with a Big Pharma often gain much-needed cash, access to Big Pharma clinical development, manufacturing and marketing strength, as well as prestige, all of which can translate into increased company value long before any products are marketed. However, when biotech companies enter into such agreements, they usually lose control of development of a drug, and must often be content with receiving upfront payments and milestone payments for research, and royalties representing a small percentage of sales of the drug, while Big Pharma pockets the lion's share of the earnings. In recent years, biotech companies have been doing deals in which they can capture a greater share of the potential profits from a drug. These especially include "risk sharing" agreements, in which the biotech company shares expenses and risks of drug development with the Big Pharma, and also shares the manufacturing, promotion and revenue opportunities.

In a growing number of cases, biotech companies needing additional cash and/or development and marketing strength to successfully develop and market their drug may do without a Big Pharma alliance altogether, thus retaining more control of their products and a greater share of the profits. One approach involves alliances between biotech companies. For example, in 2000 the Texas Biotechnology Corporation (TBC; known as Encysive Pharmaceuticals since May 2003) and Icos formed a 50:50 joint venture known as Icos-TBC, to commercialize compounds (endothelin receptor antagonists) originally developed by TBC for the treatment of congestive heart failure and other cardiovascular diseases. This limited partnership is designed to accelerate development of these compounds, to develop a commercial presence and to share in the profits of the type of large commercial market usually addressed by a Big Pharma. However, in April 2003, TBC acquired Icos' interest in the joint venture for a series of cash payments, thus regaining full control of the endothelin receptor antagonists that it originally developed.

Another approach involves a co-marketing agreement between a contract sales organization (CSO) and a biotech company, in which the CSO hires, trains and manages a sales force to market a new drug, and assumes a share of the risk (by investing in or providing an upfront loan to the biotechnology company) in return for a share of the profits. The biotechnology company in turn retains control of the drug and realizes a greater share of the profits than if it had concluded a co-marketing deal with a Big Pharma. In June 2003, Scios assumed control of the sales force created under the agreement. Innovex also had a similar agreement with CV Therapeutics to market another cardiovascular drug, Ranexa (ranolazine), which is now in preregistration with the FDA after having completed Phase III clinical trials for chronic angina. The two companies modified their agreement in July 2003 to provide complete commercialization rights for Ranexa to CV Therapeutics (including the opportunity to hire and train its own sales force). Under the modified agreement, CV Therapeutics will continue to have a commercialization services relationship with Innovex relating to Ranexa.

Biotech companies, and especially genomics companies, whose business model has been based entirely on doing collaborations with Big Pharmas, have also been changing their strategies to move toward developing drugs on their own. For example, Millennium Pharmaceuticals, whose activities once nearly exclusively involved Big Pharma collaborations, made several acquisitions to accelerate its transition into drug development. The most notable of these was the acquisition of LeukoSite in 1999 and of COR Therapeutics in 2002. Currently, both of Millennium's marketed products and all but one of its products in clinical development came from these acquisitions.

In terms of business models, biotechnology companies are of two general types: drug discovery/development companies, and tool companies – there is a good deal of overlap between the two categories. Drug discovery/development companies have the goal of developing and marketing drugs, either on their own or with partners (traditionally Big Pharmas, but now also sometimes other biotech companies). Tables 11.2 to 11.4 list examples of various types of biotechnology companies.[21]

Included in these tables are Big Biotech companies (all of which are drug discovery/development companies with marketed products that are mainly focused on recombinant DNA and/or MAb-based drugs, but which are now also developing small-molecule drugs), as well as smaller drug discovery and development biotechnology companies, and tool companies.

Table 11.2
Big Biotech Companies

Company	*Location*
Amgen	Thousand Oaks, California, USA
Genentech	South San Francisco, California, USA
Biogen*	Cambridge, Massachusetts, USA
Chiron	Emeryville, California, USA
Genzyme	Cambridge, Massachusetts, USA

Note: *See Note 20 on Biogen's merger with Idec.

Source: Haberman Associates.

Table 11.3
A Selection of Smaller Biotech Companies in Drug Discovery and Development

Company	*Location*	*Comments*
CV Therapeutics	Palo Alto, California, USA	Cardiovascular drug company; several late-stage drugs in development.
Human Genome Sciences	Rockville, Maryland, USA	Genomics company; has recombinant protein drugs in development.
Icos	Bothell, Washington, USA	Signal transduction and cellular adhesion; has an erectile dysfunction drug (in partnership with Lilly) that was launched in January 2003 in Europe and is awaiting approval by the FDA.
Ligand Pharmaceuticals	San Diego, California, USA	Biology and chemistry of steroid receptors and other nuclear receptors; signal transduction. Has marketed drugs.
Millennium Pharmaceuticals	Cambridge, Massachusetts, USA	Genomics company; has marketed and late-stage products via acquisition of LeukoSite and COR Therapeutics.
Myriad Genetics	Salt Lake City, Utah, USA	Genomics company; has marketed genetic diagnostic services, and is moving into developing small-molecule drugs.
Scios	Sunnyvale, California, USA	Primarily functional genomics and signal transduction; has a recombinant-DNA-based marketed product; acquired by Johnson & Johnson in April 2003.
Sugen	South San Francisco, California, USA	Signal transduction; acquired by Pharmacia and run as a subsidiary.
Vertex Pharmaceuticals	Cambridge, Massachusetts, USA	Structure-based drug design; has a marketed product.

Source: Haberman Associates.

Table 11.4
A Selection of Biotechnology "Tool" Companies

Company	*Location*	*Comments*
Affymetrix	Santa Clara, California, USA	Functional genomics; world's leading gene chip company.
Applied Biosystems	Foster City, California, USA	Provides advanced genomics instruments and supplies, such as DNA sequencers and mass spectrometry systems for proteomics.
ArQule	Medford, Massachusetts, USA	Combinatorial chemistry.
Celera Genomics	Rockville, Maryland, USA	Genomics; sequenced the human and several model organism genomes; has reorganized itself to become a drug discovery and development company.
CuraGen	New Haven, Connecticut, USA	Genomics, proteomics and functional genomics; transitioning into drug discovery and development.
Incyte Genomics	Palo Alto, California, USA	Genomics and proteomics; transitioning into drug discovery and development.
Large-scale Biology	Vacaville, California, USA	Proteomics.
Lexicon Genetics	The Woodlands, Texas, USA	Functional genomics; high-throughput knockout mouse technology.
LION Bioscience	Heidelberg, Germany	Bioinformatics.
Pharmacopeia	Princeton, New Jersey, USA	Combinatorial chemistry
Variagenics	Cambridge, Massachusetts, USA	Pharmacogenomics. Merged with another biotech company, Hyseq, in February 2003, to form Nuvelo, which is focusing on discovery and development of therapeutic and diagnostic products.

Note: Several notable strategic realignments or reorganizations to enable "tool" company to drug discovery/development transitions are noted in the table. However, several other companies listed in this table also have drug discovery/development operations. The primary revenues of the companies listed in this table are from marketing of technology-based systems or information and/or collaborations involving these systems.

Source: Haberman Associates.

Tool companies are technology developers, which gain revenues by marketing technology-based systems, information and/or services. Tool companies that provide technology-based services to Big Pharmas and in some cases to drug discovery/development biotechnology companies have a great deal of overlap with drug discovery/developers that develop their products in collaboration with Big Pharmas. These service providers may enter into agreements in which the Big Pharma partner provides upfront payments, milestone payments and royalties on any drugs developed by the Big Pharma using the service provider's technology.

In recent years, there has been a trend for tool companies to transform themselves into drug discovery/development companies, and especially to attempt to develop drugs on their own. The reason for this is to realize more of the value from drug development than merely receiving either payments for sales of technology products or services, or the upfront fees, milestone payments and royalties associated with a collaboration, both of which are typically a very small fraction of the revenues from a successfully marketed drug. Moreover, competition between numerous tool companies that provide similar solutions to Big Pharma R&D often results in turning the tool companies' products into low-margin commodities. This competition may also threaten the survival of a tool company. Of course, in becoming a drug development company, the tool company assumes more of the drug development risk as well.

The tool-company-to-drug-development-company transition can also be difficult to accomplish strategically and organizationally. It may involve, for example, in-licensing of the company's first drugs in development, and the acquisition of other biotech companies with technology and/or drug candidates needed for the acquiring company's strategy. For example, Celera Genomics, a leading marketer of genomic information to Big Pharmas and other drug development companies, has restructured itself to focus on drug discovery and development, acquired Axys Pharmaceuticals to gain the capacity to develop small-molecule drugs, and most recently has purchased the rights to preclinical asthma compounds from Bayer (originally developed under an Axys/Bayer collaboration). Celera's transition is still ongoing.

The tool-company-to-drug-development-company transition is another instance of the drive of biotech companies to realize more of the value of drug development, as in risk-sharing agreements, biotech–biotech deals, and co-marketing with a CSO. Successful execution of any of these strategies strengthens the position of biotech companies as competitors with Big Pharmas. In some cases, it may lead to the transformation of a biotech company to a large, profitable, fully-integrated drug development and marketing organization, as the Big Biotechs are today. This is, for example, the goal of both Millennium and Human Genome Sciences.

The rise of the biotechnology industry, its productivity, and its drive to strengthen its competitive position vis-à-vis Big Pharmas has implications for the future structure of the pharmaceutical/biotechnology industry. A June 2000 report by the Tufts Center for the Study of Drug Development reported the finding that although Big Pharma–Big Pharma mergers have reduced the numbers of large pharmaceutical companies and produced several megafirms with ever-larger R&D budgets, the number of firms that obtained FDA approval for NCEs during the 1990s has unexpectedly been increasing.[22] According to the Tufts study, the share of new drug approvals by the FDA of the four largest pharmaceutical companies declined from 30 percent in the 1960s to 18 percent in the 1990s. A similar trend was found when analyzing the top eight firms, or analyzing approvals by therapeutic class (such as oncology or cardiovascular). The number of firms obtaining approvals for NCEs increased by 84 percent from the 1970s to the 1990s. Of the 50 firms that had one NCE approved during the 1990s, 41 of them had their first-ever NCE approval. Many of the latter firms are development-stage biotechnology companies.

In terms of the structure of the pharmaceutical/biotech industry, these trends predict that although there will be a few very large pharmaceutical companies, and the Big Pharmas might consolidate further than they have already, there would also be many small innovative firms, most of which are classified as biotechnology companies. Unless the big firms can change their cultures so as to encourage innovation, and unless they develop technology strategies that enable them to use advanced drug discovery technologies more effectively, Big Pharmas are likely to become even more dependent on biotechnology companies for drug discovery than they are now. This will tend to strengthen the position of biotechnology companies and turn Big Pharmas into development, manufacturing and marketing specialists. Even some of their development and manufacturing functions may be outsourced to contract research organizations (CROs) and contract manufacturers, as is already happening today. However, Big Pharmas may alternatively acquire biotechnology companies with marketed and late-stage drugs.

Some biotechnology companies will become bigger, via acquisition of other companies and via internal development, and increase the ranks of Big Biotech. Many biotech companies, both large and small – some of them using biotech–biotech alliances and other creative leveraging strategies – will be able to develop and (depending on the therapeutic areas) market drugs without the need to partner with Big Pharmas. Biotech companies will develop many products that address markets that are too small for Big Pharmas, and they will be able to expand the market size of some of these drugs via subsequent clinical trials. Pharmacogenomics may result in the development of many more drugs that address smaller markets, and will also make at least some clinical trials cheaper by targeting groups of patients for whom the drug has a higher probability of being safe and efficacious. These factors will tend to favor biotech companies over Big Pharmas. In summary, we expect the structure of the pharmaceutical/biotechnology industry to be very dynamic, not just a progression to bigger and bigger companies via consolidation. The ability of any company or business model (even the time-honored "Big Pharma" business model itself) to thrive or even survive in this milieu is not guaranteed. The success of any company will depend on its ability to innovate and to deal with change.

The above discussion considers only the research-based pharmaceutical/ biotechnology industry, which develops novel, branded drugs. However, with the demand of society to hold down drug prices, and the increasing numbers of drugs coming off patent, another important factor in the pharmaceutical industry (both now and in the future) is the generic drug industry. In addition to generic drug companies in the US, Europe and Japan, new generic drug manufacturers and marketers are already appearing in developing countries and other parts of the world that do not host research pharmaceutical and biotechnology companies. For example, India is becoming a leading center for such companies. Some of these companies are already gearing up not only to market generic drugs in their home countries, but also to compete with generic drug companies in the United States and Europe.

Inability to Utilize Large-scale Genomics for New Drugs

Despite the large investment in genomics, proteomics and other advanced drug discovery technologies during the 1990s and early 2000s both by

Big Pharmas and by investors in biotechnology companies with large-scale genomics technology platforms, there are currently only a handful of genomics-based drugs now in clinical development. Furthermore, as discussed earlier, neither genomics nor HTS has resulted in an increase in the productivity of pharmaceutical company R&D during the 1990s and early 2000s.

Some pharmaceutical company executives and industry commentators believe that the reason for this is that these technologies have been in place only for a relatively short time. They continue to believe in the law of large numbers; that is, if a company increases the scale, throughput and efficiency of its operations, and generates hundreds of millions of data points, it will be able to identify the minute fraction of this data that will result in the blockbuster drugs of the 2000s and 2010s. For example, George M. Milne, Jr, Senior Vice President and President of Worldwide Strategic and Operations Management at Pfizer, in a July 2001 article stated that his company has achieved a projected forty-fold increase in the speed and throughput of its HTS operations, and will generate a projected 150 million data points in 2002, as compared to 3.8 million data points in 1998. He claimed that this increased efficiency and speed was already yielding better lead compounds in several therapeutic areas, and could enable Pfizer to more rapidly exploit information on targets generated by genomics research. In the same article, John Niblack, until recently president of Pfizer Global R&D, stated in 1991 that Pfizer's very large scale R&D operation would enable the company to rapidly explore the myriad hypotheses on medical applications of new molecules and mechanisms generated via genomics research.[23]

However, as discussed earlier, Dr Niblack said later that year that neither Pfizer nor other Big Pharmas (all of which have similar "bigger is better" massive-scale genomics and HTS-based drug discovery models) could generate enough drugs to sustain their growth. Many other industry leaders and industry commentators (that is, investors, analysts and consultants) have expressed disappointment at the dearth of genomics-derived drug candidates that reach the clinical stage, and even the low numbers of such candidates in preclinical studies (that is, the studies of a drug in animal models that precede entry into human clinical trials). Moreover, since the growth, and in some cases the survival, of Big Pharma companies depends on increasing the productivity of their R&D

operations in the relatively near term, many pharmaceutical industry executives are feeling the pressure to improve the results of genomics-based research.

To compound this sense of disappointment, even the acknowledged leader in moving genomics-based products into the clinic, Human Genome Sciences, has experienced significant setbacks. During the past few years, the company's two lead genomics-derived products yielded disappointing results in Phase II clinical trials. As a result, Human Genome Sciences discontinued development of one of these products, and is conducting new clinical trials of the other product after it failed in trials for two indications. The company has recently been augmenting its pipeline with non-genomics-derived products, specifically versions of existing protein drugs developed by other companies that are modified (using HGS' proprietary technology) to have better drug-delivery-related characteristics. The company's pipeline now consists of a nearly equal mix of genomics-derived and non-genomics products.

Although another acknowledged leader in genomics, Millennium Pharmaceuticals, as of October 2002 had nine drugs in clinical trials, only one of them emerged from the company's genomics-based R&D efforts. The other eight drugs came to Millennium as the result of acquisitions, as did the company's two marketed products. Two other leading genomics companies are CuraGen and Myriad Genomics. The former currently has one genomics-based therapeutic in clinical trials and the latter has none.[24]

Perhaps most people working in genomics-based drug discovery believe that the law of large numbers is not enough. For example, when researchers screen compounds against a target molecule, they attempt to determine whether any of these compounds will produce a specific effect (for example, inhibition of a target enzyme) on that target. However, if the researchers are screening against the wrong targets, and/or if they are screening libraries of compounds that do not include molecules that interact with the target, they will get meaningless answers, no matter how many compounds they screen nor how fast they do so.

In doing large-scale genomics, the problem of selecting the right targets among the very large numbers of potential targets provided by genomics is the problem of target validation. As discussed earlier, target validation is arguably the most important challenge facing genomics-based drug discovery today. Functional genomics researchers and other biologists have provided drug discovery and development companies with a host of "target validation technologies."[25] Nevertheless, drug discovery researchers remain overwhelmed by the magnitude of the target validation challenge.

The problem of obtaining the right compounds to screen against validated targets is the problem of structural diversity of chemical libraries (that is, the ability to construct libraries of compounds that are sufficiently different from each other that a small number of them will interact with any target of interest). Combinatorial chemistry allows medicinal chemists to generate large numbers of diverse compounds in chemical libraries. However, the degree of diversity in these libraries is limited by the algorithms that chemists use to determine diversity. This diversity as seen by the chemists and their algorithms may have little or nothing to do with diversity as seen by biological molecules. This is especially the case as targets being investigated by drug discovery researchers move beyond the four families of targets that are addressed by currently marketed drugs. For this reason, drug discovery researchers often refer to most of the members of this wider universe of targets as "hard targets."

Some advanced chemical technology companies, including Infinity Pharmaceuticals, NeoGenesis and Sunesis, as well as academic groups, have been developing computational methods for generating libraries that are much more diverse than conventional libraries, and as a result show evidence of being able to address a wider range of targets. For example, because of these advances, NeoGenesis has gained partnership agreements with such drug discovery and development companies as Merck, Biogen, Schering-Plough, Tularik, Pharmacia and Mitsubishi Pharmaceuticals. Other researchers have been employing structure-based drug design and other computational methods based on protein structure determination in order to design more specific drugs for targets of interest. As structural biology and structural genomics science and technology advances, these methods are likely to be more widely used.

Effective Biology-Driven Drug Discovery Strategies in the Post-Genomic Era

Although genomics-based drug discovery and development has been floundering during the 1990s and early 2000s, there are several examples of smaller-scale, biology-based programs that have been more successful in getting promising drugs into the clinic, and in some cases onto the market. Leading examples are listed in Table 11.5.

Table 11. 5
Promising Drugs in Development from Biology-Driven Drug Discovery Programs

Drug	*Company*	*Mechanism*	*Disease*	*Stage*
Glivec/Gleevec	Novartis	Inhibits Bcr-Abl and related signaling kinases	Cancer	Marketed (CML and gastrointestinal stromal tumors)
Velcade (bortezomib)	Millennium	Proteasome inhibitor	Cancer	Approved, US, May 2003 (multiple myeloma); in clinical trials in several other cancers
MLN518	Millennium	FLT-3 receptor kinase inhibitor	Acute myeloid leukemia	Phase I
PKC412	Novartis	FLT-3 receptor kinase inhibitor	Acute myeloid leukemia	Phase II
CEP-701	Cephalon	FLT-3 receptor kinase inhibitor	Acute myeloid leukemia	Phase II
SU11248	Sugen/ Pharmacia	FLT-3 receptor kinase inhibitor	Acute myeloid leukemia	Phase I
SCIO-469	Scios/Johnson & Johnson	p38 MAP kinase inhibitor	Rheumatoid arthritis	Phase II
VX-702	Vertex	p38 MAP kinase inhibitor	Acute coronary syndromes	Phase I

Source: Haberman Associates.

Perhaps the most notable of these programs is the development of Novartis' cancer drug known as Gleevec in the US.[26] Gleevec was discovered as the result of biological research beginning in the 1970s in several academic institutions, combined with medicinal chemistry studies at Novartis in the 1990s. It targets a specific molecule, known as Bcr-Abl kinase, that has been shown to cause chronic myeloid leukemia (CML).[27]

Gleevec had the fastest time to approval in the United States of any cancer drug to date – approximately 35 months from the first dose of the drug in human clinical trials to FDA approval. Since the drug's initial approval, the FDA has expanded its indications – from treatment of advanced chronic myeloid leukemia (CML), or of earlier-stage CML after other therapies had failed, to treatment of gastrointestinal stromal tumors. As the result of further positive clinical trials, further expansions of the drug's indications are expected. After 13 months on the market, worldwide sales of the drug were US$425 million.

Included in Table 11.5 are several still newer examples of promising cancer drugs discovered via biology-based research which are now in development. These include Millennium's Velcade (bortezomib), for which the company submitted an NDA to the FDA for treatment of multiple myeloma in January 2003. Velcade was approved by the FDA for that indication in May 2003. It is in clinical trials for other types of cancer as well. Velcade was discovered as the result of studying the biology of the proteasome, an enzyme complex that is involved in controlling regulatory signals within the cell that control cell proliferation and survival. This biological research was combined with medicinal chemistry by researchers at LeukoSite to discover the compound. Millennium acquired LeukoSite in 1999, and thus inherited Velcade, which owes nothing to Millennium's genomics technology programs.

Several companies – Millennium, Novartis, Cephalon and Sugen (a biotechnology company that became a division of Pharmacia as the result of an acquisition) – are targeting a mutated form of the FLT3 receptor kinase. This mutated signal transduction kinase is present on the surface of cells of a type of acute myeloid leukemia (AML) with a poor prognosis, and is thought to be critically involved in the growth and survival of these malignant cells. Drugs in this class thus represent other examples of "designer cancer drugs." These are drugs like Gleevec, designed to modulate a very well validated target that is central to the causation of a very specific type of cancer. As with Gleevec and Velcade, these compounds were developed by combining medicinal chemistry and studies of the biology of the FLT3 receptor in AML.

Other biology-led drug discovery programs are being carried out by several biotechnology companies that specialize in studying signal transduction pathways. Companies in this class that have such compounds in clinical trials include Celgene, Ligand, Metabolex, Scios, Sugen, Tularik and Vertex.[28] Several biotechnology companies that have drug discovery programs based on studies of signal transduction have been successful in bringing drugs to market, but not necessarily from their signal transduction programs. For example, Ligand, Scios and Vertex (which mainly focuses on structure-based drug design) all have marketed products.

Traditional biological research tends to be slow, as, for example, the thirty years of academic research behind Gleevec. Therefore, many companies are utilizing high-throughput advanced drug discovery technologies in order to speed up and extend the range of biological research. Unlike massive genomics-based research programs based on the law of large numbers, these programs start with a disease model, a pathway, or a biological hypothesis. The researcher then uses high-throughput technologies to study all the genes and/or proteins involved in these models, pathways or hypotheses at once, in contrast to the traditional, slow one-molecule-at-a-time approach.

For example, Scios has implemented a biology-driven multidimensional research program, which includes the application of gene chip-based gene expression profiling, advanced medicinal chemistry and other advanced drug discovery technologies to the discovery of new drugs that target signaling kinases involved in cardiovascular disease. Scios researchers start with a disease model – specifically animal models of congestive heart failure – and also examine human tissue samples from patients with this disease in parallel to their animal studies. Using gene chips, they screen for kinases and other molecules whose expression changes as the result of the disease condition. This broad program has so far resulted in one compound in Phase II clinical trials and another class of compounds in preclinical studies. Researchers in other companies and academic laboratories are also using proteomics to speed up the identification of molecules involved in a signal transduction pathway. Such pathway determination done via the methods of traditional biological research (usually in several unconnected academic laboratories) may take a decade or more.

The above type of biology-led R&D programs may utilize some of the same high-throughput technologies as do large-scale, technology-driven, genomics-based drug discovery programs. However, since they start with a specific disease model, pathway or hypothesis, they are usually much lower scale than the latter type of program. Since biology-based research strategies often connect molecular targets to a specific disease condition early in the research process, they often greatly simplify target validation.

In addition, they can also facilitate the transition from drug discovery to animal models to clinical trials.

Some leaders of Big Pharma, Big Biotech and genomics companies have been beginning to pay more attention to or to utilize biology-led drug discovery strategies. For example, according to James C. Mullen, President and CEO of Biogen, his company is utilizing a "biomining" or "gene pull" strategy, starting with a disease model, for utilizing genomics information in drug discovery.[29] Millennium appointed Julian Adams, the scientist (originally from LeukoSite) who led the effort to discover Velcade and then move it into the clinic, as Senior Vice President of Drug Discovery in 1991.[30] Also, as discussed in a later section, Novartis is building a new research laboratory in Cambridge, Massachusetts, with an organization that can facilitate biology-driven post-genomic drug discovery strategies. With the ongoing successes of biology-led research strategies and the enormity of the target validation problem, companies would do well to implement such strategies.

Organizational Challenges to Genomics-Based Drug Development

The barriers to effective drug discovery in the post-genomics era include both technical/scientific and organizational dimensions. We have discussed technical and scientific issues in the previous sections. We next turn to organizational challenges.

There are many intertwined organizational and associated strategic issues that can either facilitate or hinder effective genomics-based drug discovery and development, especially in Big Pharma companies.[31] A key problem is that these companies have tended to maintain the relatively risk-adverse bureaucratic cultures that served them well when traditional drug discovery strategies were sufficient to fill their pipelines and produce sufficient numbers of high-valued drugs to maintain their growth. These bureaucratic, risk-adverse cultures have been hindering innovation by in-house Big Pharma researchers, and have also been hampering the success of partnerships with biotech companies, which usually have more informal, innovative and risk-friendly cultures. This cultural difference is a major reason why biotech companies tend to be more productive in terms of drug discovery than Big Pharmas, and why biotech companies have thus become the main discovery engine of the industry as a whole.

Big Pharmas have an even greater mismatch in size, culture and motivation with academic groups than they have with biotech companies. As the results of academic research – especially research with implications for target validation, disease biology and advanced chemical technologies – have become more and more important for effective drug discovery in the post-genomic era, this has also hampered Big Pharma's efforts to utilize its genomics technology effectively.

Big Pharma mega-mergers also tend to make companies even more risk-averse. The reorganizations that result from these mergers also often disrupt internal R&D efforts and partnerships with biotech companies and with academia. For example, during such a reorganization, R&D staff experience uncertainties about whether they will continue to be employed by the newly merged company and to whom they will report. This uncertainty can paralyze R&D activity for a year or more.

Moving Towards More Effective Drug Discovery and Development Organizations

Several Big Pharmas, realizing the need to move toward more innovative R&D organizations and cultures, are developing various strategies. One of these involves restructuring R&D programs to attempt to create a more innovation-friendly environment. All companies that have concluded a merger must reorganize their R&D organizations in order to integrate operations from the two merging companies and to eliminate redundancies. However, these reorganizations (as well as frequent reorganizations in non-merged companies) are in most cases unfavorable for R&D innovation. Nevertheless, Big Pharma executives often attempt to use restructuring to improve their companies' R&D output.

Perhaps the most direct attempt to use reorganization to create an innovation-friendly environment is that being undertaken by Glaxo-SmithKline (GSK).[32] In 2001, following the merger between Glaxo Wellcome and SmithKline Beecham that created GSK, the company set up six Centers of Excellence for Drug Discovery (CEDDs). These units, each centered on a particular therapeutic area or group of therapeutic areas (such as neurology, respiratory disease and rheumatoid arthritis), work at the lead optimization and preclinical stage of drug development. GSKs massive technology-driven genomics and HTS platform provides the CEDDs with lead compounds. Scientists in the CEDDs then optimize the leads, conduct preclinical testing, weed out compounds deemed likely to fail, and prepare compounds that emerge from their organizations for clinical trials. GSK's unified corporate clinical development unit then takes over to complete the development of the drug.

Each CEDD contains between two hundred and eighty and four hundred scientists. Within each CEDD, the scientists work in a similar way to a biotech company, with very little bureaucracy. Moreover, the scientists receive incentives, including cash bonuses and stock options, tied to successful development of a new drug.

GSK thus is attempting to combine the advantages of a huge megafirm with those of an entrepreneurial biotech company. However, earlier-stage drug discovery is still run by the technology-driven "law of large numbers" approach. (The CEDDs can, however, also test and optimize compounds developed by external partner companies, presumably including compounds discovered via biology-based research.) As with all recent Big Pharma reorganization plans, the success of the GSK approach remains to be seen.

Another approach to developing more innovative Big Pharma R&D organizations is to establish satellite laboratories near academic centers of excellence. For example, several Big Pharmas have selected the Boston area, and especially the area of Cambridge, Massachusetts, near the Massachusetts Institute of Technology (MIT), as the site of their satellite laboratories. This area of Cambridge is also the site of numerous biotech companies.

Big Pharma or Big Biotech companies that have or are building satellite laboratories in the Boston area include Amgen, AstraZeneca, Merck and Pfizer. These satellite laboratories are designed to function like small, innovative biotech companies. Being located near academic centers of excellence, they can recruit young PhDs with excellent credentials, and put them to work under an experienced research director. They can also interact with academic scientists and, especially in the MIT area, with scientists in small biotech companies.

Satellite laboratories usually focus on a particular area of special interest to the parent company. For example, the Pfizer Discovery Technology Center in Cambridge focuses on the integration of genomics, biological sciences, advanced chemical technologies, engineering and informatics, with a special emphasis on discovering and designing libraries of compounds that interact with particular protein families. Center researchers, for example, collaborate with the nearby Medford-based combinatorial chemistry company ArQule in chemogenomics research to screen compound libraries against protein families.

The other function of satellite laboratories is to catalyze change in the parent R&D organization, in order to help it become more innovation-friendly. This is facilitated by keeping the satellite laboratory independent, giving it a strong leader with authority in the parent R&D organization, establishing strong lines of communication with departments in the larger company, and structuring the company's R&D units so that they can implement any approaches developed by the satellite laboratory. The key to catalyzing change in the larger organization, as well as effectively utilizing any technology, targets and compounds coming out of the satellite laboratory, is to manage the process of moving people and projects between the satellite laboratory and the parent R&D organization. Once again, the verdict is still awaited on the issue of how well satellite laboratories work in enabling Big Pharma R&D to become more innovative and actually improve its output of new drugs.

A third, and more radical approach to creating a more innovative Big Pharma R&D organization is being taken by one company, Novartis. This approach is to create a new company R&D headquarters in an academic center of excellence. In May 2002, Novartis announced that it was creating the Novartis Institute of Biomedical Research (NIBR) to be located in Cambridge, Massachusetts; once again near MIT. The company's worldwide R&D operations in the US, Europe and Japan will be led from the new Cambridge headquarters.

In another radical departure from usual Big Pharma practices, Novartis chose an academic, Mark Fishman MD, to head up the NIBR. He is a former Professor of Medicine at Harvard Medical School, and Chief of Cardiology and Director of Cardiovascular Research at Massachusetts General Hospital in Boston. Dr Fishman's field of research has been focused on the zebra fish as a model organism, and especially its use in understanding the developmental genetics of the cardiovascular system. Dr Fishman expects to infuse the NIBR with the spirit of cutting-edge academic research, including collaboration between the fields of genomics, genetics, biology, chemistry and medicine. He also intends to break down the usual Big Pharma culture of secrecy, and encourage Novartis scientists to attend scientific meetings, publish in scientific journals, and to consult and collaborate with outside scientists. He believes that creating this melding of industry and academic cultures may well take years, as will the generation of new streams of drugs resulting from such an approach. In this mission, Dr Fishman has the full backing of Novartis CEO Daniel Vasella, who personally recruited Dr Fishman.[33]

The fact that Dr Vasella went for such a radical and long-term approach to the organization of a Big Pharma R&D operation shows that at least one Big Pharma CEO realizes the depth of the need for organizational change and organizational learning if Big Pharmas are to meet the challenges of the post-genomics era.

Conclusions

Meeting the Challenges of the 2000–10 Decade and Beyond

If the research-based pharmaceutical industry – or any one company in that industry – is to have a future, it must implement science and technology strategies that will enable it to utilize the new biotechnologies much more effectively than it is currently doing, in order to develop the new drugs that it will need for its own growth and survival and to realize the tremendous opportunities that the new science and technologies are expected to provide. As we have discussed, the need for such strategic change has both a scientific/technological and an organizational dimension.

In addition to its internal issues, the pharmaceutical industry is facing and will continue to face many pressures from society. The foremost issue in the news is the demand to control drug prices. However, new breakthrough drugs have the potential to lower costs due to hospitalization, surgery and other aspects of healthcare, which may more than offset the fact that breakthrough drugs are bound to be expensive due to their high development costs. New breakthrough drugs and other types of therapeutics expected in the future (such as cellular therapies, tissue-engineering implants and organ replacements, and gene therapies), including treatments or even cures for now intractable diseases, will also improve the quality of life of many of the world's people. The future availability of such breakthrough therapeutics will depend on continuing negotiation of the social contract between patients, physicians, national health services, insurance companies, and pharmaceutical and biotechnology companies.

Another social trend that is expected to become more important in the next decades is the aging of the population in many parts of the world. Aging patients and others afflicted with major diseases, their families, and those who expect to be afflicted in the future, are demanding better treatments and even cures. News of scientific advances and of the development of breakthrough drugs such as Gleevec, is fueling these demands. These social trends go counter to the demand for controlling drug prices, and need to be reconciled with them.

When the pharmaceutical industry learns to utilize its technologies effectively, it will be better able to meet the needs of society, and thus ensure its own future.

Appendix

Advanced Drug Discovery Technologies

Allan B. Haberman

The principal advanced drug discovery technologies referred to in Chapter 11, "Biotechnology and the Future of the Pharmaceutical Industry" are briefly described below:

Genomics

Genomics forms the core of drug discovery technologies and indicates the study of the genetic information of an organism. (The entire complement of genes in an organism is known as its genome.) The rapid development of genomics began with the 1990 launch of the international Human Genome Project, under the leadership of the US National Institutes of Health (NIH) and Department of Energy (DOE). In the late 1990s, a commercial company, Celera Genomics, was launched with the goal of completing the sequence of the human genome faster than the "public" Human Genome Project, and making the information available to its paying subscribers. As a result of this competition, both the "public" and "private" human genome sequencing efforts completed a working draft sequence of the human genome simultaneously, and several years ahead of schedule, in 2000. A "finished" genome sequence was completed in April 2003.

In addition, genome sequences for many microorganisms (including important pathogens) have been determined; these sequences are expected to enable researchers to develop new generations of antibiotics, vaccines, and other agents for the treatment and prevention of infectious diseases. Researchers have also determined genome sequences for such model organisms as the yeast *Saccharomyces cerevisiae*, the nematode worm *Caenorhabditis elegans*, the fruit fly *Drosophila melanogaster*, and the laboratory mouse. Ongoing model organism genome sequencing projects include the genomes of the laboratory rat and the zebra fish, a vertebrate model organism increasingly used in developmental biology studies.

Genomic information from model organisms is useful in helping drug discovery researchers to determine the function of human genes and in developing experimental disease models. Experimental studies with these disease models, as well as other types of genomics, functional genomics and biological studies, help researchers identify which molecules may be centrally involved in a disease process. If researchers can design drugs that modulate these molecules, these may be good lead candidates for development as human therapeutics. A biomolecule (usually, but not always, a protein) that interacts with a drug or drug candidate is known as a molecular target.

Target validation

Target validation may be defined as determining that a molecular target is critically involved in a disease process and a potentially valuable point of intervention for new drugs. Since genomics has provided drug discovery researchers with an enormous glut of potential new targets, target validation is arguably the most important challenge facing genomics-based drug discovery today.

Proteomics

This term refers to the study of the entire protein complement of an organism, which is known as the proteome. Proteomics focuses on studying protein expression in cells and tissues, as well as protein function, with a particular focus on interactions between proteins. Knowledge of protein expression is important, because many physiological processes, including those important in disease, occur at the level of protein expression, modification and degradation rather than at the level of gene expression. Moreover, an understanding of the ways that proteins interact with each other enables researchers to gain knowledge of physiological pathways, and to identify potential molecular targets for drug discovery. Proteomics is much less developed as a field than is genomics, both because technologies used in proteomics are less developed and because of the greater complexity of the proteome and of protein–protein interactions as compared to the genome.

Functional Genomics and Functional Proteomics

Functional genomics refers to a group of platform technologies aimed at the large-scale systematic, and often genome-wide determination of gene function. Large-scale systematic determination of protein function, known as functional proteomics includes, for example, methods for the systematic study of protein–protein interactions and of protein modifications that occur during cell signaling. Functional proteomics as a discipline overlaps with functional genomics. Functional genomics has evolved because genomics has presented researchers with an enormous number of genes and the proteins that they encode, for which little is known about their biology and function. In particular, drug discovery researchers need this functional information for the identification and validation of drug targets.

Functional genomics platform technologies encompass a wide variety of methods. Perhaps the best known is microarray or "gene chip" technology. In this technology, gene chip companies or researchers themselves construct arrays of DNA probes on glass or silicon surfaces that resemble a computer chip. These arrays can be used to monitor the expression of large numbers of genes (up to all the genes in an organism's genome) in various cells and tissues, and how they vary during disease processes and drug treatment.

Other functional genomics methods include knockout mouse technology, in which mice with deletions (or "knockouts") of single genes are produced in order to study the function of the genes. Lexicon Genetics has developed high-throughput methods for systematic production of knockout mice, potentially for all genes in the mouse genome. The knockout strategy, including systematic production of single-gene knockouts, has also been extended to other model organisms, such as yeast and *C. elegans*.

Other functional genomics technologies include the use of model organisms to study the function of human genes, and high-throughput studies of gene function (including the effects of drugs on gene or protein expression or on the localization of key proteins within a cell) in cultured cells.

Combinatorial Chemistry

This involves synthesizing large libraries of chemical compounds, instead of the traditional one-molecule-at-a-time method of organic synthesis. The resulting compounds are usually small synthetic organic molecules (such as those that are traditionally developed as drug candidates in the pharmaceutical industry), but can also be peptides, oligonucleotides or oligosaccharides. In the case of the production of libraries of small organic molecules, combinatorial chemistry generally involves three basic steps:

- Systematic methods for chemically modifying a starting compound.
- Automated synthesis of large numbers of compounds using these methods.
- A means of indexing or keeping track of the large numbers of compounds created by these methods.

High-Throughput Screening

With libraries of compounds (either created by combinatorial chemistry, or, for example, libraries of natural products or the traditional compound files of a pharmaceutical company), automated methods are used to systematically assay these compounds for a desired biological activity, in a high-throughput manner. This process is known as high-throughput screening (HTS). Assays for use in HTS may be traditional types of assays such as enzyme- or receptor-binding assays, or assays developed via functional genomics technologies (for example, cellular assays with a fluorescent read-out).

Structure-Based Drug Design

Structure-based drug design consists of computer-augmented methods for using the three-dimensional structure of a target protein to design more specific drugs. Such methods, as well as other advanced chemical technologies, enable medicinal chemists to synthesize compounds that may be more effective drug leads than merely using combinatorial chemistry based on known "drug-like" molecules. These methods may also help chemists to design drugs that modulate types of target molecules beyond the small number of families of targets addressed by traditional medicinal chemistry.

Bioinformatics

This term refers to computer analysis of biological data. It involves storage, handling, making accessible and facilitating analysis of data from such areas as genomics, proteomics and combinatorial chemistry. Cheminformatics refers to the computer analysis of data from advanced chemical technologies as well as related information needed for drug discovery and basic research, such as patient data. Bioinformatics is being driven by the enormous amount of data that is accumulating in large databases of DNA and protein sequences, as well as other types of information, such as information on protein–protein interactions, three-dimensional structures of proteins and effects of drug interactions on these structures, and information on gene expression (for instance, derived from gene chips) and protein expression in various cells, tissues and organisms under different conditions.

Bioinformatics is essential for working with data from genomics and other advanced drug discovery technologies. For example, various informatics tools have been developed to aid in mapping and sequencing of DNA, and to access, search and analyze DNA sequence data. One important function of these tools is to identify the coding regions of genes and to search for sequence similarities. Researchers may use similarity search algorithms, for example, to identify human genes that are homologous to the gene for a target of an existing small-molecule drug on the market. Such genes are likely to encode proteins that belong to the same protein family as the known target. If any of these novel proteins can be validated as targets for particular diseases, they are likely to be modulated by small organic molecules that are similar to the known compounds (such as the original marketed drug) that modulate the original known target. The researchers can then screen combinatorial libraries of compounds that are similar to the original marketed drug, and thus find lead drug candidates for the novel targets. Screening of combinatorial libraries of compounds against entire families of proteins, and using protein structure information to design "biased libraries" that contain compounds that are more likely to interact with proteins in a particular family of interest, is referred to as "chemogenomics" by some researchers and companies.

As genomics, proteomics and other advanced drug discovery technologies advance, they present increasingly difficult challenges for bioinformatics and computational biology. For example, researchers are developing *in silico* models for understanding biological pathways, and models of microbial cells that are designed to predict, for example, the effects of specific drugs on these cells. On a more immediately practical level, pharmaceutical and biotechnology companies would like to have data and knowledge-management systems that integrate all the functions and departments of genomics-based R&D. Such systems would need to integrate genomics, proteomics and functional genomics, with chemical, pharmacological, biological and perhaps clinical data. This type of system would facilitate the ability of researchers to move from genomics and proteomics data to validated targets and lead compounds. These systems are difficult to develop because of incompatibilities of software between different disciplines and different types of data (for example, sequence data versus images), and even between software used to handle the same type of data in different laboratories. Bioinformatics companies (notably LION Bioscience and Ipedo) have developed the first systems designed to integrate different drug discovery disciplines. However, these are in a very early stage of development.

Pharmacogenomics

Pharmacogenomics refers to the study of how genetic differences between individuals affect their response to drugs, and therefore affect the outcome of drug treatment. Genetic differences between individuals can affect how a drug is metabolized. For example, it may be difficult to give a large enough dose to be therapeutically effective in individuals who are fast metabolizers of a particular drug. Conversely, levels of the drug may build up to such high levels in slow metabolizers that the drug reaches toxic levels. If physicians knew the "fast metabolizer" or "slow metabolizer" status of their patients, they could adjust the dose of the drug accordingly. Genetic differences between patients can also affect the toxicity and efficacy of drugs in other ways.

The embryonic science of pharmacogenomics is made possible by the development of rapid, high-throughput means of discovering common genetic differences between individuals, and for mapping these genetic variations. The most common type of genetic variation is called a single nucleotide polymorphism (SNP). Genetic differences between individuals that occur at the same location on a chromosome are referred to as polymorphisms; SNPs are differences in a single nucleotide base. Other genetic polymorphisms include insertions, deletions, duplications and rearrangements of genetic material; these polymorphisms involve greater amounts of DNA information than do SNPs.

Once SNPs have been determined and mapped, researchers may perform "genotyping" of an individual – that is, determining which of a large number of SNPs is contained in his or her genome. If researchers have determined that certain of these SNPs are associated with response to specific drugs or with disease, a person's SNP genotype may help physicians predict that person's response to a specific drug.

Applications of Pharmocogenomics

Pharmacogenomics researchers and other industry experts project that pharmacogenomics will lead to an age of "personalized medicine," in which drugs are designed for populations with specific genetic markers. In this scenario, pharmaceutical and biotechnology companies will develop these drugs together with specific genetic diagnostic tests that will pinpoint whether a patient belongs to the population group for which a specific drug was designed. Physicians will then administer the diagnostic test to the patient before prescribing the drug.

However, the application of pharmacogenomics in drug discovery and development is in its infancy. SNP analysis and other genetic methods are widely used in target discovery. This involves identifying populations in which a particular SNP or group of SNPs results in disease, and mapping the SNPs to a particular protein-coding region of a gene, which then is deemed to code for a target protein. However, pharmacogenomics is not yet widely utilized in other aspects of drug discovery and development.

Pharmacogenomics is finding use in drug discovery and early-stage drug development, in order to eliminate compounds that may be toxic as early as possible in the drug development process. It is also much more utilized in cancer than in other therapeutic areas. Cancer arises as the result of genetic changes that occur in certain cells of the body, as opposed to genetic variation in the germline of the individual. As these changes accumulate, they enable these cells to escape from normal physiological controls, to become cancerous, and eventually to metastasize. Researchers are attempting to design drugs that are targeted to genetic changes in specific types of cancer cells, which are usually more extensive than single-nucleotide mutations. Another application of pharmacogenomics in cancer is identifying patients for whom certain existing chemotherapeutic drugs are likely to result in severe side effects. Moreover, there are two new cancer drugs now on the market that must be prescribed only after conducting diagnostic tests to show that a patient's cancer cells express the target molecules that these drugs inhibit. However the diagnostic tests used in prescribing these drugs involve either fluorescent in situ hybridization (FISH) (staining of chromosomes with a fluorescent labeled DNA probe that is specific for the abnormal target genes, and visualizing the results under a microscope) or immunohistochemistry (staining of cells with a fluorescent antibody that is specific for the target protein, and visualizing the results under the microscope), not SNP analysis or other genotyping methods.

However, in other therapeutic areas, pharmacogenomics is applied much less often. There are concerns in the industry that designing drugs for small genetically defined groups of patients does not fit the usual business model of Big Pharma, which aims at developing billion-dollar blockbuster drugs that can treat large groups of patients. However, many currently marketed drugs do not work well for large percentages of the patient populations for which they are prescribed, either in terms of efficacy or in terms of side effects. Patients and/or their physicians who are dissatisfied with the results of drug therapy may discontinue taking the drug, thus reducing revenues to the pharmaceutical company that developed it. (This is especially an issue with drugs designed for long-term or lifelong use, such as the cholesterol-lowering statin drugs, drugs for high blood pressure, or drugs for type 2 diabetes.) A more targeted drug may have a higher rate of compliance, at least partially offsetting the smaller population that it is indicated for and the need to develop multiple drugs for different populations of patients with the same disease.

Nevertheless, there are major disincentives to develop pharmacogenomic tests for patient responses to drugs already on the market, since such tests would reduce utilization of these drugs. Pharmacogenomics is more likely to be applied to cancer, where the consequences of using the wrong drug for a given patient are often life threatening. Moreover, as discussed earlier, drugs targeted to small markets (in this case, genetically defined subpopulations of patients with a common disease) may be more likely to be developed by smaller companies rather than the largest of the Big Pharmas.

However, in the long term, as the science of pharmacogenomics advances and results in applications with large markets, the age of personalized medicine will be upon us. For example, drugs may eventually be developed that can prevent the appearance of a disease (such as type 2 diabetes or certain forms of cancer) in genetically susceptible populations. Such drugs would be taken over a lifetime, resulting in large revenues for their developers. In the nearer term, development of numerous genetically targeted drugs for smaller populations (perhaps mainly by smaller companies) to treat rather than prevent specific diseases may also become common during the 2010–20 decade.

CONTRIBUTORS

DR. JAMAL S. AL-SUWAIDI is Director General of the Emirates Center for Strategic Studies and Research (ECSSR) and Professor at the UAE University. He has taught courses in political methodology, political culture, comparative governments and international relations at the UAE University and the University of Wisconsin. He earned his MA and PhD degrees in Political Science at the University of Wisconsin.

Dr. Al-Suwaidi is the author of numerous articles on a variety of topics, including perceptions of democracy in Arab and Western societies, women and development, and UAE public opinion on the Gulf crisis. He is the author of "Gulf Security and the Iranian Challenge," *Security Dialogue* vol. 27, no. 3 (1996); a contributing author to *Democracy, War and Peace in the Middle East*; *Education and the Arab World: Challenges of the Next Millennium* (ECSSR, 1999); *The Gulf: Future Security and British Policy* (ECSSR, 2000); *Leadership and Management in the Information Age* (ECSSR, 2002) and *Human Resource Development in a Knowledge-Based Economy* (ECSSR, 2003). Dr. Al-Suwaidi is the editor of *The Yemeni War of 1994: Causes and Consequences* (ECSSR, 1996) and *The Gulf Cooperation Council: Prospects for the Twenty-first Century* (ECSSR, 1999). He is the editor and contributing author of *Iran and the Gulf: A Search for Stability* (ECSSR, 1996), published both in English and Arabic. ECSSR won two awards for this book: Best Publisher and Best Book in the field of Humanities and Social Sciences. Dr. Al-Suwaidi is also the co-editor of *Air/Missile Defense, Counterproliferation and Security Policy Planning* (ECSSR, 1999).

DR. SUE BAILEY was US Assistant Secretary of Defense for Health Affairs from 1998 to 2000, and Deputy Assistant Secretary of Defense for Clinical Services from 1994 to 1995. Prior to her appointment as Deputy Assistant Secretary of Defense for Clinical Services, Dr. Bailey attained the rank of Lieutenant Commander. She served as a spokesperson for former President Bill Clinton's Healthcare Reform Campaign in 1993. She is presently a news analyst for NBC. As Assistant Secretary of Defense for Health Affairs, Dr. Bailey was responsible for the health protection of the United States military forces, including pre- and post-deployment health, battlefield casualties, endemic disease and environmental hazards. During her tenure, she traveled to war zones and locations of engagement including Guantanamo Bay, Haiti, the Arabian Gulf, Bosnia and Kosovo.

Dr. Bailey is a board-certified psychiatrist whose clinical and academic background includes a faculty position at Georgetown University Medical School. Proficient in sign language, she served as consultant to Gallaudet College for the Deaf. She has been an advocate for the legal and medical needs of the disabled community, has published and consulted on a variety of healthcare subjects, and is a recognized expert in her field. Dr. Bailey's professional education includes a BS degree from the University of Maryland and a DO degree from Philadelphia College of Osteopathic Medicine. She completed her internship and served in residency at George Washington University and completed a Medical Post-Graduate Fellowship at Johns Hopkins University. Dr. Bailey also completed additional specialty courses in Emergency Medicine at Howard University, Internal Medicine at Harvard Medical School, and Psychiatry and Neurology at Yale University School of Medicine.

MR. JUAN ENRIQUEZ-CABOT is a Senior Research Fellow and Director of the Harvard Business School Life Science Project. His field of research is located at the intersection of science, economics and public policy. He focuses on the interplay between technology, economics, politics and the rise of the life sciences, and analyzes how computers, genomics and other new technologies are shaping the present and future. His previous positions include CEO of Mexico City's Urban Development Corporation, Coordinator General of Economic Policy, and Chief of Staff for Mexico's Secretary of State. Juan Enriquez has held a number of positions at Harvard, including Fellow at the Center for International Affairs. He serves on the Board of the US State Department Advisory Committee on Economic Policy (Biotechnology Group), the Genetics Advisory Council of the Harvard Medical School, the Board of Cabot Microelectronics, the Chairman's International Council of the Americas Society, and The Institute for Genomic Research.

Juan Enriquez has written numerous Harvard Business School case studies as well as articles for various publications including *Science*, *Foreign Policy*, *The New York Times*, and *Trends in Biotechnology*. He is contributing editor of *The Journal of Biolaw and Business*. Juan Enriquez's article "Transforming Life, Transforming Business" (2000, with Ray Goldberg), expected to have a major influence on the actions of business managers worldwide, won a 2000 McKinsey Award. Juan Enriquez is the author of *As the Future Catches You: How Genomics and Other Forces are Changing Your Life, Your Work, Your Investments, Your World* (2001).

DR. JOHN D. GEARHART is the C. Michael Armstrong Professor of Medicine at the Institute of Cell Engineering at Johns Hopkins University. He is also Professor of Gynecology and Obstetrics, Physiology and Comparative Medicine at the Johns Hopkins University School of Medicine, as well as Professor of Biochemistry and Molecular Biology at the Johns Hopkins University's Bloomberg School of Public Health. A leading scientist in the field of genetic engineering of cells, Dr. Gearhart's work has focused on the development and use of human reproductive technologies. His highly significant research on the isolation and study of human embryonic stem cells has paved the way for the development of tissue-transplantation therapies for degenerative diseases and injuries such as diabetes, Parkinson's disease, stroke and spinal cord injuries. He led the research team responsible for first identifying and isolating human pluripotent stem cells. The ground-breaking results were published in the much quoted report titled "New Potential for Human Embryonic Stem Cells" (John D. Gearhart et al., *Science* vol. 282, November 6, 1998). He has authored or co-authored as many as 221 publications relating to transgenesis, Down's syndrome and stem cells.

Dr. Gearhart holds a BS degree from the Pennsylvania State University, a Master's degree in Plant Genetics from the University of New Hampshire, and a doctoral degree in Genetics from Cornell University. He has received several scientific honors including the Basil O'Connor Starter Research Award from the March of Dimes Birth Defects Foundation, and the Golden Plate Award from the American Academy of Achievement. He was selected as a Joseph P. Kennedy Jr Scholar in Mental Retardation.

DR. RAY A. GOLDBERG is Professor Emeritus of Agriculture and Business at Harvard Business School, where he served as Faculty Chairman of the Agribusiness Senior Management Seminar. He is Chairman of the Advisory Panel for the World Bank *Guide to Developing Agricultural Markets and Agro-Enterprises*. Dr. Goldberg also served on the National

Research Council as Chairman of the Subcommittee on Economic and Social Development in a Global Context, a subdivision of the Board on Agriculture and National Resources. Dr. Goldberg has served on the boards of directors of over forty major agribusiness firms, farm cooperatives and technology firms. He is one of the founders and first President of the International Agribusiness Management Association and an advisor and consultant to numerous government agencies, financial institutions and private firms. Dr. Goldberg has been Professor at the Harvard Graduate School of Business Administration since 1955 and became Professor Emeritus in July 1997. He is an Honorary Professor and a Member of the Royal Agricultural College at Cirencester, England.

Dr. Goldberg is the author, co-author and editor of 23 books, and has published 110 articles. He has also authored and supervised the development of over 1,000 case studies. His most recent publications involve developing strategies for private, public and cooperative managers as they position their firms, institutions and government agencies in a rapidly changing global food system. He also conducts research on the major biological, logistical, packaging and informational revolutions that affect global agribusiness. Recent articles include "The Business of Agroceuticals" in *Nature Biotechnology* (Supplement to vol. 17, 1999) and "Food Wars: A Potential Peace" in *The Journal of Law, Medicine and Ethics* (Supplement to vol. 28, no. 4, 2000). The article "Transforming Life, Transforming Business: The Life-Science Revolution" in the *Harvard Business Review* (March–April 2000, with Juan Enriquez), received a 2000 McKinsey Award.

MR. ANDREW GREENE is Chief Operating Officer of Merlin Biosciences, where he oversees investment, reporting and fundraising activities. In October 2002 he became Merlin's Chief Executive, responsible for all aspects of the firm relating to the Financial Services Authority of the United Kingdom. As the Senior Investment Director, he is a member of the Investment Committee and of Merlin Biosciences Board of Directors. He is a Director of portfolio companies of the Merlin Fund L.P. and the Merlin Biosciences Fund L.P. Mr. Greene joined Merlin from Merrill Lynch International, where he was Director of Healthcare Investment Banking. Prior to that, he founded the Healthcare Investment Banking team at HSBC Investment Bank. Mr. Greene was also a senior member of Cap Gemini's Life Sciences strategic consulting group, where he ran major projects for Pharmacia, Eli Lilly, Glaxo and SmithKline Beecham. He spent ten years with the international management consulting firm, Booz, Allen & Hamilton, Inc. in its Chicago and London offices, during which time he was head of the European Healthcare practice, and a senior member of the US healthcare practice.

Mr. Greene has been an adviser to the British National Health Service. He served as an expert on the Healthcare 2000 panel to devise a funding strategy for the National Health Service. He also advised the Health Ministers of the previous government. Before joining Booz, Allen & Hamilton, Mr. Greene worked in Corporate Treasury for the US hospital chain Lifemark Corporation in Houston, Texas. Mr. Greene also worked as a technician on a US$10 million National Institutes of Health study at Harvard Medical School/Massachusetts General Hospital in Boston. He has a Master's degree in business administration and accounting from Rice University and a Bachelor's degree in history from Cornell University. He is the author of numerous articles and treatises on healthcare policy.

DR. ALLAN B. HABERMAN is the Principal of Haberman Associates, a ten-year-old consulting firm specializing in science and technology strategy for pharmaceutical, biotechnology and life science companies. He is also a Principal and Founder of the Biopharmaceutical Consortium, an expert team formed in 1997 to assist life science companies, research groups and emerging enterprises to identify and exploit promising breakthrough technologies. Dr. Haberman's consulting engagements include work on new product development and technology strategy, opportunity assessment, assessment of drug pipelines, partnering, and due diligence. Prior to forming Haberman Associates, Dr. Haberman was the Associate Director of the Biotechnology Engineering Center at Tufts University. He received his PhD in biochemistry and molecular biology from Harvard University.

Dr. Haberman is the author of numerous publications on the pharmaceutical and biotechnology industries, their technologies and products, and on the major therapeutic areas for drug discovery and development. His more recent publications include *Biochemical Pathway and System Analysis for Target Identification and Validation* (2002); *Target Identification and Validation: Key Approaches for Improving the Efficiency and Profitability of Drug Discovery and Development* (2001, with Deirdre Lockwood and Malorye Branca); *Next-Generation Metabolic Disease Therapeutics: An*

Analysis of Eight Therapeutic Pipelines for Diabetes and Obesity (2001) and *From Data to Drugs: Strategies for Benefiting from New Drug Discovery Technologies* (1999, with George Lenz and Dennis Vaccaro).

DR. MICHIO KAKU is the Henry Semat Professor of Theoretical Physics at the City College and the Graduate Center of the City University of New York. Previously he was Visiting Professor at the Institute of Advanced Study at Princeton University and at New York University and has lectured around the world. His research interest is directed at developing a "theory of everything," and thus helping to complete the work of Albert Einstein to capture the laws of the physical universe in one brief formula. Dr. Kaku is the co-developer of string field theory. He also wrote the first paper on conformal supergravity and the breakdown of supersymmetry at high temperatures. He is an internationally recognized authority on theoretical physics as well as on the environment. Dr. Kaku graduated summa cum laude from Harvard University in 1968. He received his PhD from the University of California at Berkeley in 1972.

Dr. Kaku's textbooks are required reading at many of the most renowned physics laboratories. His books *Hyperspace: A Scientific Odyssey Through Parallel Universes, Time Warps, and the 10th Dimension* (1994) and *Visions: How Science Will Revolutionize the 21st Century* (1997) became international bestsellers, and have been translated into several different languages. Other publications include *To Win a Nuclear War* (1986, with Daniel Axelrod); *Beyond Einstein: Superstrings and the Quest for the Final Theory* (1995, with Jennifer Trainer); and *Strings, Conformal Fields and M Theory* (2000). Dr. Kaku has appeared on CNN, PBS, the BBC and other broadcasting stations, as well as on over six hundred radio programs around the United States, and has been quoted by leading international newspapers such as *The Washington Post*, *The New York Times*, and *The Times*.

DR. GLENN MCGEE is Professor of Bioethics, Philosophy and History as well as Professor of the Sociology of Science at the Department of Bioethics at the University of Pennsylvania School of Medicine. He is also a Senior Fellow at the Center for Bioethics. His most widely read book is *The Perfect Baby: A Pragmatic Approach to Genetics* (1997, 2nd edn 2000), the first articulation of pragmatic theory of bioethics and a frequently cited treatment of ethical issues in reproductive genetics. Dr.

McGee obtained his PhD from Vanderbilt University in 1994. He is a member of the Philosophy of Medicine Committee of the American Philosophical Association and the Society for the Advancement of American Philosophy.

Dr. McGee has served on numerous federal genetics task forces and as ethicist to the evaluation panel for genetic tests at the US Food and Drug Administration. He has consulted with patient advocacy groups, candidates for national office and corporate boards and testified before the US House of Representatives and the Senate. Dr. McGee and cloning scientist Ian Wilmut co-authored the widely cited "Model for Regulating Human Cloning" in *The Human Cloning Debate*, which he edited in 1998 (2nd edn 2000). Dr. McGee has co-authored the first analysis of stem cell patenting for a book that he also co-edited, entitled *Who Owns Life?* (2002). Other publications include *Pragmatic Bioethics* (1999, 2nd edn 2002), which he edited, and the forthcoming *Generation Genome*. He has also authored more than two hundred articles and essays and edited five books. Dr. McGee is editor-in-chief of *The American Journal of Bioethics*, as well as Senior Editor of The MIT Press' bioethics book series *Basic Bioethics*.

DR. JOHN PIERCE is Director of Biochemical Sciences and Engineering at DuPont Central Research and Development laboratories in Wilmington, Delaware. He previously served as Director of Strategic Planning and Resources and as Director of Genetic Technology at DuPont Agricultural Products, where he was responsible for research efforts in agricultural biotechnology and its applications in the food, feed and materials markets. Dr. Pierce specializes in research and product development for agricultural and industrial uses. With over forty publications in scientific journals, he is frequently quoted in the press and addresses international audiences. He joined DuPont in 1982 and currently leads DuPont's research into the use of biotechnology in the design and production of industrially important materials. With a BS in Biochemistry from Pennsylvania State University and a PhD in Biochemistry from Michigan State University, Dr. Pierce worked as a Research Fellow at the University of Wisconsin and also at Cornell University. From 1981 to 1982, he served as a National Science Foundation Postdoctoral Fellow and as a National Institutes of Health Postdoctoral Fellow in 1982.

Dr. Pierce's publications include "Cloning of Higher Plant Omega-3 Fatty Acid Desaturates" (co-authored) in *Plant Physiology* (vol. 103, 1993); "Using Biotechnology to Enhance and Safeguard the Food Supply: Delivering the Benefits of the Technology" in *World Food Security and Sustainability: The Impacts of Biotechnology and Industrial Consolidation* (1999, National Agricultural Biotechnology Council report), and "Realizing the Promise of Plant and Microbial Biotechnology" in *The Journal of BioLaw and Business* (Supplement 2000, US Department of State).

MS. HELEN QUIGLEY received her MA degree in Middle East and Islamic Studies from the University of Toronto, Canada. She graduated from McGill University in the same subject. Ms. Quigley spent two and a half years working in the Middle East Department of Harvard College Library, where she worked on a small collection of Turkish language materials written in the Armenian script as well as helping to produce a publishable version of the Library's Armenian catalogue. She was a researcher for *A Reference Guide to Modern Armenian Literature, 1500–1920* by K.B. Bardakjian and in the past, her research has focused on twelfth- to thirteenth-century Turkish history. For the past twelve years she has been at the Tozzer Library of Harvard University, cataloguing foreign language materials as well as conducting an inventory of the rare books held in the library, which includes codices, Mayan language materials and archaelogical/ethnographic field notes.

MR. JEREMY RIFKIN is the Founder and President of the Foundation on Economic Trends in Washington, DC. He has been a Fellow at the Wharton School of Finance and Commerce of the University of Pennsylvania since 1994, where he lectures in the executive and advanced management programs. His work focuses on the impact of scientific and technological changes on the economy, the workforce, society and the environment. His books have been translated into more than twenty languages and are used in universities around the world. His best-selling book *The End of Work* (1995) is widely credited with helping to shape the current debate on technology displacement, corporate downsizing and the future of jobs. His 1998 international bestseller *The Biotech Century: Harnessing the Gene and Remaking the World* addresses the many critical issues accompanying the new era of genetic commerce and is the most widely read book in the world in this field. Mr. Rifkin's 2000 bestseller, *The Age of Access*, explores the vast changes occurring in the capitalist system as it makes the transition from geographic markets to e-commerce networks and from industrial to cultural production. In his newest book *The Hydrogen Economy*, published in September 2002, Rifkin takes us on an eye-opening journey into the next great commercial era in history. He envisions the dawn of a new economy powered by hydrogen that will fundamentally change the nature of our market, political and social institutions, just as coal and steam power did at the beginning of the industrial age.

Mr. Rifkin has appeared frequently on television programs, including CNN Crossfire, ABC Nightline and Larry King Live and is a regular commentator for *The Guardian*, *Sueddeutsche Zeitung*, *Le Monde*, *El Pais*, *Panorama* and the *Los Angeles Times*. Mr. Rifkin holds a degree in economics from the Wharton School of Finance and Commerce of the University of Pennsylvania, and a degree in International Affairs from the Fletcher School of Law and Diplomacy at Tufts University. He is a consultant to heads of state and government officials around the world and speaks frequently before business, labor and civic forums.

DR. GREGORY STOCK is Director of the Program on Medicine, Technology and Society at the University of California at Los Angeles School of Public Health and a visiting professor in the Department of Health Services. He is also visiting Senior Fellow at the Center for the Study of Evolution and the Origin of Life. His research focuses on the Human Genome Project and its spin-offs in reproductive biology, genetic testing, drug development, healthcare policy and medical practice. Dr. Stock has published research papers on developmental biology, limb regeneration, and laser light scattering, and has designed computer software for electronic banking networks. In his book *Metaman: The Merging of Humans and Machines into a Global Superorganism* (1993), Dr. Stock has explored the larger evolutionary significance of humanity's recent technological progress. Following the publication of *Metaman*, he spent a year as a Fellow at the Woodrow Wilson School of Public and International Affairs at Princeton University. Dr. Stock holds a PhD in Biophysics from Johns Hopkins University and an MBA from Harvard University.

Dr. Stock's numerous publications on medical and ethical aspects of the genomics revolution include *The Book of Questions* (1985), which became a *New York Times* bestseller and has been translated into seventeen languages. Other publications include *Engineering the Human Germline: An Exploration of the Science and Ethics of Altering the Genes We Pass to Our Children* (1999, with John Campbell, eds) and *Redesigning Humans: Our Inevitable Genetic Future* (2002).

Notes

Introduction

1 Richard W. Oliver, *The Coming Biotech Age: The Business of Bio-Materials* (New York, NY: McGraw-Hill, 2000), 7.
2 Jeremy Rifkin, *The Biotech Century: The Coming Age of Genetic Commerce* (London: Victor Gollancz, 1998, and New York, NY: Jeremy Tarcher, Inc., 1998), xii.
3 Ibid.
4 Ibid., 234.
5 Stan Davis and Bill Davidson, *2020 Vision: Transform Your Business Today to Succeed in Tomorrow's Economy* (New York, NY: Fireside, 1991), 187.
6 Thomas C. Wiegele, *Biotechnology and International Relations: The Political Dimensions* (Gainesville, FL: University of Florida Press, 1991), 1.
7 Scott P. Layne, Tony J. Beugelsdijk and C. Kumar N. Patel (eds), *Firepower in the Lab: Automation in the Fight Against Infectious Diseases and Bioterrorism* (Washington, DC: Joseph Henry Press, 2001), 267.
8 Rifkin, *The Biotech Century*, 3.
9 Wiegele, *Biotechnology and International Relations*, 9.
10 Oliver, *The Coming Biotech Age*, 56.
11 Ibid., 2.
12 Ibid., 221.
13 Ibid., 48.
14 E. Balázs, E. Galante, J.M. Lynch et al. (eds), *Biological Resource Management: Connecting Science and Policy* (Berlin, Heidelberg: Springer-Verlag, 2000), 188.
15 Ibid., 179.
16 Oliver, *The Coming Biotech Age*, 142.
17 Rifkin, *The Biotech Century*, xii.
18 Oliver, *The Coming Biotech Age*, 41, 45, 47.
19 Rifkin, *The Biotech Century*, 195.
20 Ibid.
21 Ibid.
22 Oliver, *The Coming Biotech Age*, 96.
23 Ibid., 106.
24 Ibid., 238.
25 Ibid.
26 Caldwell (1987) in Wiegele, *Biotechnology and International Relations*, 6.
27 Oliver, *The Coming Biotech Age*, 49.
28 Balázs et al. (eds), *Biological Resource Management, 188.*
29 Ibid.

Chapter 1

1 "New Cure for Bubble Boy Disease," April 17, 2002, CBS News (www.cbs.com); see also *New England Journal of Medicine*, April 2002.
2 "First Gene Therapy for Alzheimer's," ABC News, April 11, 2001 (abcnews.go.com/sections/living/DailyNews/alzheimersgenetherapy010411.html).
3 Interview with Dr Leonard Hayflick. See also Leonard Hayflick, *How and Why We Age* (New York, NY: Ballantine Books, 1994).
4 Michio Kaku, *Visions: How Science will Revolutionize the 21st Century* (New York, NY: Anchor Books, 1997) 211; See also *New York Times*, December 7, 1993.
5 See *Science* 279 (1998) 349–352; See also *Nature Genetics*, January 1999; "Immortalizing Enzyme Does Not Make Human Cells Cancerous," University of Texas Southwestern Medical Center, January 4, 1999.
6 "New Study Shows Normal-Looking Clones May Be Abnormal," Whitehead Institute for Biomedical Research, MIT, Cambridge, MA, July 9, 2001 (www.wi.mit.edu).
7 "Scientists Decipher the Genetic Code of Malaria," The Institute for Genomic Research, Rockville, Maryland, October 2, 2002 (TIGR.org). See also *Nature* magazine, October 2002.
8 "Technique Could Improve Cartilage Repair," MIT News, July 18, 2002 (Thomson@mit.edu).
9 Kaku, *Visions*, 159. See also *Scientific American*, September 1996, 42.
10 "A Better Breed of Fish?: FDA Petitioned to Halt Release of Bio-engineered Salmon," ABC News, May 11, 2001.
11 Kaku, *Visions*, 218; Interview with W. Gilbert.
12 "Raising the Mammoth," Discovery Channel, 2000 (www.discovery.com).

Chapter 2

1 Julie A. Nordlee, Steve L. Taylor, Jeffrey A. Townsend, Laurie A. Thomas and Robert K. Bush, "Identification of a Brazil-Nut Allergen in Transgenic Soybeans," *New England Journal of Medicine*, March 14, 1996, 688.
2 Marion Nestle, "Allergies to Transgenic Foods: Questions of Policy," *New England Journal of Medicine*, March 14, 1996, 726.
3 Aldous Huxley, *Brave New World* (Garden City, NY: Doubleday, Doran and Company, Inc., 1932).
4 Burke E. Zimmerman, "Human Germ-Line Therapy: The Case for its Development and Use," *Journal of Medicine and Philosophy*, December 1991, 594.
5 "Whether to Make Perfect Humans," *New York Times*, July 22, 1982, A22.

6 Jon Beckwith, "A Historical View of Social Responsibility in Genetics," *BioScience*, May 1993, 330.

Chapter 3

1 This chapter refers mainly to Arabic-speaking countries in the Middle East although some data and comparisons also include Islamic states from North Africa, Europe and Asia.
2 Bernard Lewis, *What Went Wrong? Western Impact and Middle Eastern Response* (Oxford: Oxford University Press, 2002) 3, 6.
3 Jared Diamond, *Germs, Guns, and Steel: The Fates of Human Societies* (New York, NY: W.W. Norton), 267.
4 Juan Enriquez, "Too Many Flags?" *Foreign Policy* (Fall, 1999).
5 Recent advances in genomics allow each of us to trace back our ancestry to a very few individuals in Europe and even fewer, earlier, individuals in Africa. See for example, Bryan Sykes, *Seven Daughters of Eve* (New York, NY: W.W. Norton, 2001) or Richard Dawkins, *A River Out of Eden* (London: Basic Books, reprint edition, 1996).
6 Steve Olson. *Mapping Human History: Discovering Our Past Through Our Genes* (New York, NY: Houghton Mifflin, 2002).
7 Angus Maddison, *The World Economy* (Paris: OECD, 2001).
8 Ibid., 46.
9 Jared Diamond, *Germs, Guns, and Steel*, 141.
10 Figure 3.3, "Worldwide Oil Production (1000 b/d)," is taken from *International Petroleum Encyclopedia, 2002*, 211–215. Figure 3.4, "Oil Production as a Percentage of World Gross Production," is taken from World Bank, *World Development Indicators, 2002* (CD ROM). *International Petroleum Encyclopedia, 2002, Basic Petroleum Data Book, Petroleum Industry Statistics,* vol. XXII, no. 2, August 2002. The price used for this calculation was the annual average wellhead price of united crude oil.
11 Nicholas Negroponte, *Being Digital* (New York, NY: Vintage Books, 1996).
12 Varian Hal and Peter Lyman, *How Much Information?* December 2000 (http://www.press.umich.edu/jep/06-02/lyman.html).
13 World Bank, *World Development Indicators 2002* (CD ROM).
14 Bruce R. Scott and George C. Lodge (eds), *US Competitiveness in the World Economy* (Boston: Harvard Business School Press, 1985).
15 James Watson and Francis Crick, "A Structure for Deoxyribose Nucleic Acid," *Nature* Vol. 171, April 25, 1953.
16 Juan Enriquez and Rodrigo Martinez, *The Biotechonomy (1.0): A Rough Map of Biodata Information Flow*, Harvard Business School Working Paper (Winter 2002) 03–028.

17 See www.nexia.com.
18 See www.google.com.
19 Genetics Advisory Council Meetings, HMS, 2002.
20 Enriquez and Martinez, *The Biotechonomy.*
21 See http://www.ncbi.nlm.nih.gov/Genbank/GenbankOverview.html or http://www.ebi.ac.uk/embl/ or http://www.ddbj.nig.ac.jp.
22 Otto Solbrig, Robert Paarlberg and Francesco Di Castri. *Globalization and the Rural Environment* (Cambridge, MA: Harvard University Press, 2001).
23 Juan Enriquez and Ray Goldberg. "Transforming Life, Transforming Business: The Life Science Revolution," *Harvard Business Review* (April/May 2000).
24 Keith Bradsher, "Vietnam Becomes a Safe Haven for Tourists," *The New York Times,* International section, January 5, 2003.
25 Dick Teresi, *Lost Discoveries: Ancient Roots of Modern Science from the Babylonians to the Maya* (New York, NY: Simon & Schuster, 2002). Seyyed Hossein Nasr, *Islamic Science: An Illustrated Study* (World of Islam Festival Publishing Co, 1976). UNESCO International Exhibition on Islamic Science and Technology (www.unesco.org/pao/ exhib/islam.htm). Malaysia's Science Center, in Kuala Lumpur, also hosted a show on Islamic science during 2002.
26 One of the great early treatises on surgery in Europe, comprising four titles in one volume is *Cyrurgia parva Guidonis. Cyrurgia Albucasis cum cauteriis et aliis instrumentis. Tratatus de oculis Jesui Hali. Tratatus de oculis Canamusali.* Authored by Guy de Chauliac (circa 1300-1368), and published in Venice, 1500, this work is based on the work of 10th and 11th century Islamic physicians and surgeons such as Abu'l Qasim Al-Zahrawi and Ali Ibn-Isa Al Kahhal.
27 See http://www.nlm.nih.gov/
28 Jared Diamond, *Germs, Guns, and Steel: The Fates of Human Societies* (New York, NY: W.W. Norton), 247.
29 See Lewis, op. cit.
30 See Nabil Matar (ed. and trans.) *In the Lands of the Christians: Arabic Travel Writing in the Seventeenth Century* (New York, NY: Routledge, 2003).
31 *Oxford Encyclopedia of the Modern Islamic World* Vol. 4 (New York, NY: Oxford University Press, 1995), 15.
32 Angus Maddison. *Monitoring the World Economy 1820-1992* (Paris: OECD, 2000), 19.
33 The World Bank, *2002 Development Indicators*, Table 6.5.
34 The World Bank, *2002 Development Indicators*. Table 2.5.
35 The World Bank, *2002 Development Indicators*. Table 2.13.
36 Bruce R. Scott, "The Great Divide in the Global Village," *Foreign Affairs* 80, no.1 (January 2001): 160-177.

37 Based on data from Angus Maddison, *Monitoring the World Economy 1820-1992.* (Paris: OECD, 2000) Tables F-4, D-1b, C16.
38 The World Bank, *2002 Development Indicators,* Table 5.11.
39 Reproduced in Niels Bohr, *Atomic Physics and Human Knowledge* (New York, NY: John Wiley & Sons, 1958).
40 Erwin Schrödinger, *What is Life? The Physical Aspect of the Living Cell* (Cambridge, Cambridge University Press, 1948).
41 See http://www.bu.edu/remotesensing/Faculty/El-Baz/FEBcv.htm
42 There are many debates as to why it is hard to attract and retain scholars in the area. For one example of this debate see Dr. Shafeeq N. Ghabra, Director of Strategic and Future Studies, Kuwait University. (http://www.inthenationalinterest.com/Articles/Vol2Issue1/Vol2Issue1Ghabra.html)
43 Islamic countries included for comparison are Algeria, Bangladesh, Benin, Burkina Faso, Cameroon, Chad, Ivory Coast, Egypt, Gabon, Gambia, Guinea-Bissau, Guyana, Indonesia, Iran, Jordan, Kuwait, Libya, Malaysia, Mali, Mauritania, Morocco, Niger, Nigeria, Oman, Pakistan, Qatar, Saudi Arabia, Senegal, Sierra Leone, Sudan, Suriname, Syria, Togo, Tunisia, Turkey, Uganda and United Arab Emirates.
44 Quoted in *Islamic Science: An Illustrated Study*, page 8.

Chapter 4

1 Maureen Dowd, "From Botox to Botulism," *New York Times,* Opinion-Editorial, September 25, 2001.
2 J.L. Resnick, L.S. Bixler, L. Cheng and P.J. Donovan, "Long-term Proliferation of Mouse Primordial Germ Cells in Culture." *Nature* 359 (1992): 550–551; See also Y. Matsui, K. Zsebo and B.L. Hogan, "Derivation of Pluripotential Embryonic Stem Cells from Murine Primordial Germ Cells in Culture," *Cell* 70 (1992): 841–847.
3 Michael J. Shamblott, Joyce Axelman, Shunping Wang, Elizabeth M. Bugg, John W. Littlefield, Peter J. Donovan, Paul D. Blumenthal, George R. Huggins and John D. Gearhart, "Derivation of Pluripotent Stem Cells from Cultured Human Primordial Germ Cells," *Proceedings of the National Academy of Sciences, USA* 95 (1998): 13726–13731.
4 POU (Pit-Oct-Unc) is an acronym for a family of protein factors that bind DNA and permit the DNA to be transcribed into RNA. Pit-Oct-Unc were the first three members of the family to be identified. Oct4 or Oct3/4 is a member of the POU family of transcription factors. It is short for Octamer-4 protein (or factor), which is the fourth member of the Oct group – a group of proteins that bind eight bases of DNA.

5 H. Spemann (1943) "Forschung und Leben" (Research and Life), as quoted in T.J. Horder, J.A. Witkowski and C.C. Wylie, *A History of Embryology* (Cambridge: Cambridge University Press, 1986), 219.
6 J.W. McDonald, X.Z. Liu, Y. Qu, S. Liu, S.K. Mickey, D. Turetsky, D.I. Gottlieb and D.W. Choi, "Transplanted Embryonic Stem Cells Survive, Differentiate and Promote Recovery in Injured Rat Spinal Cord," *Nature Medicine* 5 (1999): 1410–1412.
7 Lars M. Björklund, Rosario Sánchez-Pernaute, Sangmi Chung, Therese Andersson, Iris Yin Ching Chen, Kevin St. P. McNaught, Anna-Liisa Brownell, Bruce G. Jenkins, Claes Wahlestedt, Kwang-Soo Kim and Ole Isacson, "Embryonic Stem Cells Develop into Functional Dopaminergic Neurons after Transplantation in a Parkinson Rat Model," *Proceedings of the National Academy of Sciences USA* 99 (2002): 2344–2349; Jong-Hoon Kim, Jonathan M. Auerbach, José A. Rodríquez-Gómez, Iván Velasco, Denise Gavin, Nadya Lumelsky, Sang-Hun Lee, John Nguyen, Rosario Sánchez-Pernaute, Krys Bankiewicz and Ron McKay, "Dopamine Neurons Derived from Embryonic Stem Cells Function in an Animal Model of Parkinson's Disease," *Nature* 418 (2002): 50–56.
8 William M. Rideout III, Konrad Hochedlinger, Michael Kyba, George Q. Daley and Rudolf Jaenisch, "Correction of a Genetic Defect by Nuclear Transplantation and Combined Cell and Gene Therapy," *Cell* 109 (2002): 17–27.
9 B. Sorio, E. Roche, G. Berna, T. Leon-Quinto, J.A. Reig, F. Martin, "Insulin-secreting Cells Derived from Embryonic Stem Cells Normalize Glycemia in Streptozotocin-Induced Diabetic Mice," *Diabetes* 49 (2001): 157–162; Nadya Lumelsky, Olivier Blondel, Pascal Laeng, Ivan Velasco, Rea Ravin and Ron McKay, "Differentiation of Embryonic Stem Cells to Insulin-Secreting Structures Similar to Pancreatic Islets," *Science* 292 (2001): 1389–1394.
10 R.B. Cervantes, J.R. Stringer, C. Shao, J.A. Tischfield and P.J. Stambrook, "Embryonic Stem Cells and Somatic Cells Differ in Mutation Frequency and Type." *Proceedings of the National Academy of Sciences, USA* 99 (2002): 3586–3590.
11 Paul Berg, Nobel Laureate, US Senate Testimony, March 19, 2003.
12 From a speech delivered by British Prime Minister Benjamin Disraeli on July 24, 1877. See *Bartlett's Familiar Quotations.*

Chapter 5

1 James A. Thomson, Joseph Iskovitz-Eldor, Sander S. Shapiro, Michelle A. Waknitz, Jennifer J. Swiergiel, Vivienne S. Marshall and Jeffrey M. Jones,

"Embryonic Stem Cell Lines Derived From Human Blastocysts," *Science* 282 (1998): 1145-1147; J. Gearhart, "New Potential for Human Embryonic Stem Cells," *Science* 282 (1998): 1061–1062.

2 P.R. Wolpe and G. McGee, "'Expert Bioethics' as Professional Discourse: The Case of Stem Cells," in S. Holland, K. Lebacqz and L. Zoloth (eds) *The Human Embryonic Stem Cell Debate* (Cambridge, MA: MIT Press, 2001), 204–242.

3 M. Farley, "Roman Catholic Views on Research Involving Human Embryonic Stem Cells," in Holland, S. et al., 113–119.

4 T.B. Okarma, "Human Embryonic Stem Cells: A Primer on the Technology and its Medical Applications," in S. Holland, et al., 3–14.

5 Ibid.

6 Ibid.

7 A.R. Chapman, M.S. Frankel and M.S. Garfinkel, "Stem Cell Research and Applications: Monitoring the Frontiers of Biomedical Research, *AAAS/ICS Report*, 1999.

8 National Bioethics Advisory Commission, *Ethical Issues in Human Stem Cell Research* (US Government Printing Press, 1999).

9 It was suggested that totipotent cells might be removed from 4- or 8-cell pre-implantation embryos destined for in vitro fertilization (without destroying the embryo, a technology performed with some frequency in contemporary reproductive therapeutic settings for purposes of pre-implantation genetic diagnosis); see J.W. McDonald, X.Z. Liu, Y. Qu, S. Liu, S.K. Mickey, D. Turetsky, and D.I. Gott, "Transplanted Embryonic Stem Cells Survive, Differentiate, and Promote Recovery in Injured Rat Spinal Cord." *Nature Medicine* 5 (1999): 1410–1412. It was also suggested that scientists who conduct embryonic cell research ought not themselves to be engaged in the destruction of embryos, or, perhaps more correctly, that the activity of embryonic cell research could or should be viewed as morally distinct from that of obtaining cells through the destruction of embryos. See National Bioethics Advisory Commission, Ibid.

10 President G.W. Bush, "Address to the Nation on Stem Cell Research," Vanderbilt Television News Archives, August 2001.

11 Friend, T. "Half of Stem Cell Money Could Go to Royalties," *USA Today* August 13, 2001, A1.

Chapter 8

1 See Organisation for Economic Co-operation and Development (OECD). "Industrial Sustainability and the Role of Biotechnology," in *Biotechnology*

for Clean Industrial Products and Processes – Towards Industrial Sustainability (Paris: OECD, 1998), 15–27.

2 National Research Council (NRC), “Range of Bio-Based Products,” in *Bio-Based Industrial Products – Priorities for Research and Commercialization* (Washington, DC: National Academy Press, 2000) 55–73.

3 National Research Council (NRC), “Processing Technologies,” in *Bio-Based Industrial Products*, op. cit., 74–102.

4 O. Kirk, T.V. Borchert and C.C. Fuglsang, “Industrial Enzyme Applications,” *Current Opinion in Biotechnology* vol. 13 (2002): 345–351.

5 Ibid.

6 Organisation for Economic Co-operation and Development (OECD). “Enzymatic Production of Acrylamide (Mitsubishi Rayon, Japan),” in *The Application of Biotechnology to Industrial Sustainability – Sustainable Development* (Paris: OECD, 2001), 71–76.

7 Organisation for Economic Co-operation and Development (OECD), “Manufacture of Riboflavin (Vitamin B2) (Hoffmann, La-Roche, Germany),” in *The Application of Biotechnology*, op. cit., 51–55.

8 Organisation for Economic Co-operation and Development (OECD), “Polymers from Renewable Resources (Cargill Dow, United States),” in *The Application of Biotechnology*, op. cit., 87–90.

9 National Renewable Energy Laboratory (NREL), “Bioethanol – Moving into the Marketplace,” in *Biofuels for Sustainable Transportation* (Washington, DC: NREL. 2001).

10 See www.jxj.com.

11 S.R. Fahnestock, “Fibrous Proteins from Recombinant Microorganisms,” in *Biopolymers* vol. 8, S. Fahnestock and A. Steinbuches (eds), *Polyamides and Complex Proteinaceous Materials* (Weinheim, Germany: Wiley-VCH, 2003).

12 C. Viney, A.E. Huber, D.L. Dunaway, K. Kerkam and S.T. Case, “Optical Characterization of Silk Secretions and Fibers,” in D. Kaplan, W.W. Adama, B. Farmer and C. Viney (eds) “Silk Polymers,” *ACS Symposium Series*, vol. 544, American Chemical Society, Washington, DC (1994): 120–136. Also F. Vollrath and D.P. Knight, “Liquid Crystalline Spinning of Spider Silk,” *Nature* vol. 410 (2001): 541–548. See also P.J. Willcox, S.P. Gido, W. Muller and D.L. Kaplan. “Evidence of a Cholesteric Liquid Crystalline Phase in Natural Silk Spinning Processes.” *Macromolecule*, vol. 29 (1996): 5106–5110.

13 Baeuerlein, E., *Biomineralization: From Biology to Biotechnology and Medical Applications* (New York, NY: VCH, 2001).

14 A.M. Belcher, X.C. Wu, R.J. Christensen, P.K. Hansma, G.D. Stucky and D.E. Morse, “Fidelity of $CaCO_3$ Crystal Phase-Switching and Orientation Controlled by Shell Proteins,” *Nature* vol. 381 (1996): 56–58. Also A.M. Belcher, P.K. Hansma, G.D. Stucky and D.E. Morse, “First Steps in Harnessing the Potential of Biomineralization as a Route to New High-Performance Composite Materials. *Acta Materialia* vol. 46, no. 3

(1998): 733–736. See also S. Fields, "The Interplay of Biology and Technology," *Proceedings of the National Academy of Sciences* vol. 98 (2001): 10051–10054; M. Fritz and D.E. Morse. "The Formation of Highly Organized Biogenic Polymer/Ceramic Composite Materials: The High-Performance Microlaminate of Molluscan Nacre." *Current Opinion in Colloid and Interface Science* vol. 3, no. 1 (1998): 55–62; M. Fritz, A.M. Belcher, M. Radmacher, D.A. Walters, P.K. Hansma, G.D. Stucky, D.E. Morse and S. Mann. "Flat Pearl from Biofabrication of Organized Composites on Inorganic Substrates." *Nature* 371 (1994): 49–51; M.F. Weber, C.A. Stover, L.R. Gilbert, T.J. Nevitt and A.J. Ouderkirk. "Giant Birefringent Optics in Multilayer Polymer Mirrors," *Science* 287 (2001): 2451–2456.

15 See Belcher et al., 1996, op. cit.

16 See Belcher et al., 1998, op. cit.

17 K. Shimizu, J. Cha, G.D. Stucky and D.E. Morse, "Silicatein α: Cathepsin L-like Protein in Sponge Biosilica," *Proceedings of the National Academy of Sciences*, Washington, DC, vol. 95, no. 11 (1998): 6234–6238.

18 See S. Fields, 2001, op. cit.

19 See M.F. Weber et al., 2001, op. cit.

20 P. Forrer, S. Jung and A. Pluckthun, "Beyond Binding: Using Phage Display to Select for Structure, Folding and Enzymatic Activity in Proteins," *Current Opinion in Structural Biology* vol. 9, no. 4 (1999): 514–20; also R.W. Roberts and W.W. Ja. "In Vitro Selection of Nucleic Acids and Proteins: What are we Learning?" *Current Opinion in Structural Biology* vol. 9, no. 4 (1999): 521–529.

21 S.R. Whaley, D.S. English, E.L. Hu, P.F. Barbara and A.M. Belcher, "Selection of Peptides with Semiconductor Binding Specificity for Directed Nanocrystal Assembly," *Nature* vol. 405, no. 6787 (2000): 665–668.

Chapter 9

1 Dr Gordon Conway, President of the Rockefeller Foundation, as quoted in *The Scientist*, vol. 14, no. 21, October 30, 2000, 10.

2 The two arms of the United Nations – The Food and Agriculture Organization (FAO) and the World Health Organization (WHO) agreed to jointly sponsor a Codex Alimentarius (Latin for "food code") to improve food quality and safety around the world. The FAO/WHO formally established the Codex Alimentarius Commission in 1962 in Rome to oversee the effort. The Codex Commission is an international body, with membership that is open to any member or associate of the FAO or WHO. The Commission has 167 members representing 98 percent of the world's population. It meets every two years and reports directly to the Directors-General of the FAO and WHO.

3 See report issued on January 14, 2003 by the International Service for the Acquisition of Agribiotech Applications (ISAAA).
4 The Global Conservation Trust has been created by a partnership between the United Nations Food and Agriculture Organization (FAO) and the sixteen Future Harvest Centers of the Consultative Group on International Agricultural Research (CGIAR). The goal of the Global Conservation Trust is to support conservation of crop diversity in the long term.

Chapter 10

1 G. Stock, *Redesigning Humans: Our Inevitable Genetic Future* (Boston, MA: Houghton Mifflin, 2002) 41.
2 The broad consequences of this were discussed at the landmark symposium that I moderated at UCLA in January 2003. "The Storefront Genome" was organized by the Center for Society, the Individual and Genetics (http://www.arc2.ucla.edu/csig/symp1.htm).
3 See for example http://www.aber.ac.uk/~mpgwww/Metabol/Metabol.html
4 W. Grody, "Cystic Fibrosis: Molecular Diagnosis, Population Screening, and Public Policy," *Archives of Pathology and Laboratory Medicine* 123 (1999): 1041–1046.
5 G. Kolata, "Pushing Limits of the Human Lifespan," *New York Times*, March 9, 1999, page D1.
6 See, for example, http://thomas.loc.gov/cgi-bin/query/z?c108:S.245: this is the Brownback Bill introduced in the United States Senate in early 2003.
7 G. Steen, *DNA and Destiny: Nature and Nurture in Human Behavior* (New York, NY: Plenum, 1996), 113–135.
8 D. Macer et al., "International Perceptions and Approval of Gene Therapy," *Human Gene Therapy* 6 (1995): 791–803.
9 G. Stock, *Redesigning Humans*, Chapter 4.

Chapter 11

1 A "blockbuster" drug is one that generates over US$1 billion per year in peak revenues.
2 In this chapter, the trade name of a drug is used if it has one, and the generic name is given in parentheses.
3 A. Michaels, "Pfizer R&D Unable to Sustain Group Growth Rate," *Financial Times*, September 12, 2001, 30.
4 On January 28, 2003 at the World Economic Forum in Davos, Switzerland, Pfizer CEO Henry McKinnell stated that Pfizer expects, following the

expected closure of its merger with Pharmacia, to generate the new drugs it needs over the next five years from its own internal pipeline and from partnerships, (Michael Shields, Reuters, January 28, 2003).

5 The FDA defines a "new molecular entity" (NME) as a medication containing an active substance that has never before been approved for marketing in any form in the United States.

6 B.M. Bolten and T. DeGregorio, "Trends in Development Cycles," *Nature Reviews Drug Discovery* vol. 1 (2002): 335–336.

7 However, Lilly had two drug approvals in the US and one in Europe late in 2002, despite continuing manufacturing quality concerns.

8 The same month, Schering-Plough received FDA approval to begin selling Claritin over the counter, rather than by prescription, in the United States.

9 The two companies are continuing to develop the drug, and working with the FDA to ensure an optimal clinical development program.

10 Tufts Center for the Study of Drug Development, press release, November 30, 2001 (http://csdd.tufts.edu/).

11 Until September 2002, two separate divisions of the FDA reviewed small-molecule drugs and biopharmaceuticals (mainly protein drugs made via recombinant DNA and/or monoclonal antibody technologies). Review of both types of drugs was placed under one division, the Center for Drug Evaluation and Research, in September 2002. The biotechnology industry expects that this reorganization will further speed the approval process for biopharmaceuticals.

12 K.I. Kaitin and E.M. Healy, "The New Drug Approvals of 1996, 1997 and 1998: Drug Development Trends in the User Fee Era," *Drug Information Journal* vol. 34 (2000): 1–14.

13 For example, as Roche has done with Genentech.

14 As in the case of Johnson & Johnson with Centocor, and Wyeth with the former Genetics Institute.

15 J. Drews, "Drug Discovery: A Historical Perspective," *Science* vol. 287 (2000): 1960–1964.

16 Ibid.

17 One example of recombinant human insulin is Lilly's Humulin. Human growth hormone is manufactured by Genentech, Novo Nordisk, Lilly and Pharmacia. Examples of factor VIII are Bayer's Kogenate and Wyeth's ReFacto.

18 Examples of recombinant erythropoietin include Amgen's Epogen, for treatment of anemia in chronic renal failure patients on dialysis. Interferon products for the treatment of multiple sclerosis include Berlex's betaseron (interferon beta-1b) and Biogen's Avonex (interferon beta-1a).

19 For an excellent glossary of technical terms used in genomics and other advanced drug discovery technology (including those used in this chapter)

see the online Genomics Glossary hosted by Cambridge Healthtech Institute (http://www.genomicglossaries.com/).

20 In June 2003, Biogen announced that is was merging with Idec to form Biogen Idec Inc., which will be headquartered in Cambridge, Massachusetts. The two companies expect the merger to be completed in the second or third quarter of 2003.

21 There are many more biotechnology companies than those listed in Tables 11.2 and 11.3. The selection of companies for these tables is somewhat arbitrary, and is meant to provide examples of the types of existing companies.

22 K.I. Kaitin, "Pharmaceutical Industry Innovation is More Dispersed Despite M&A Activity," *Tufts Center for the Study of Drug Development Impact Report* vol. 2, no. 5 (June 2000).

23 J.K. Borchardt, "The Business of Pharmacogenomics," *Modern Drug Discovery* vol. 4 (2001): 35–39.

24 However, Myriad markets several genomics-based diagnostic tests, and all three companies have several genomics-derived drugs in preclinical studies.

25 A.B. Haberman, D. Lockwood and M.A. Branca, *Target Identification and Validation: Key Approaches for Improving the Efficiency and Profitability of Drug Discovery and Development*, CHI Reports, Cambridge Healthtech Institute, June 30, 2001. See also A.B. Haberman, *Biochemical Pathway and System Analysis for Target Identification and Validation*, CHI Reports, Cambridge Healthtech Institute, April 24, 2002.

26 Gleevec is imatinib mesylate, also known as STI 571. It is known as Glivec in countries other than the US.

27 A kinase is an enzyme that transfers phosphate groups to other molecules. Kinases that phosphorylate certain proteins are key molecules in signal transduction pathways. Bcr-Abl is an abnormal form of such a signaling kinase.

28 See Haberman, 2002, op. cit. Bcr-Abl and FLT3 are signaling molecules as well, and thus the drugs that target these kinases, as discussed above, also target signal transduction pathways.

29 See Biogen's press release of January 9, 2001 (http://www.biogen.com).

30 In September 2003, Dr. Julian Adams left Millennium to become Chief Scientific Officer at Infinity Pharmaceuticals.

31 For a full discussion of these issues, see A.B. Haberman, G.R. Lenz and D.E. Vaccaro, *From Data to Drugs: Strategies for Benefiting from New Drug Discovery Technologies*, Scrip Reports, Morpace Pharma Group and PJB Publications Ltd, July 1999.

32 S. Warner, "Science in Small Hubs," *The Scientist* vol. 16, no. 7 (April 1, 2002): 54.

33 N. Aoki, "Off the Academic Path," *Boston Globe*, May 12, 2002.

BIBLIOGRAPHY

ABC News. "First Gene Therapy for Alzheimer's." April 11, 2001. (abcnews.go.com/sections/living/DailyNews/alzheimersgenetherapy010411.html).

ABC News. "A Better Breed of Fish? FDA Petitioned to Halt Release of Bio-engineered Salmon," May 11, 2001.

Amar, Zohar. *The History of Medicine in Jerusalem* (Oxford: Archaeopress, 2002).

Aoki, N. "Off the Academic Path." *Boston Globe*, May 12, 2002.

Baeuerlein, E. *Biomineralization: From Biology to Biotechnology and Medical Applications* (New York, NY: VCH, 2001).

Balázs, E., E. Galante, J.M. Lynch et al. (eds). *Biological Resource Management: Connecting Science and Policy* (Berlin, Heidelberg: Springer-Verlag, 2000).

Beckwith, Jon. "A Historical View of Social Responsibility in Genetics," *BioScience*, May 1993.

Belcher, A.M., P.K. Hansma, G.D. Stucky and D.E. Morse. "First Steps in Harnessing the Potential of Biomineralization as a Route to New High-Performance Composite Materials. *Acta Materialia* vol. 46, no. 3 (1998): 733–736.

Belcher, A.M., X.C. Wu, R.J. Christensen, P.K. Hansma, G.D. Stucky and D.E. Morse. "Fidelity of $CaCO_3$ Crystal Phase-Switching and Orientation Controlled by Shell Proteins." *Nature* vol. 381 (1996): 56–58.

Biogen press release, January 9, 2001 (http://www.biogen.com).

Björklund, Lars M. et al. "Embryonic Stem Cells Develop into Functional Dopaminergic Neurons after Transplantation in a Parkinson Rat Model." *Proceedings of the National Academy of Sciences USA* 99 (2002): 2344–2349.

Bloom, Jonathan M. *Paper Before Print: The History and Impact of Paper in the Islamic World* (New Haven, CT: Yale University, 2001).

Bohr, Niels. *Atomic Physics and Human Knowledge* (New York, NY: John Wiley & Sons, 1958).

Bolten, B.M. and T. DeGregorio. "Trends in Development Cycles." *Nature Reviews Drug Discovery* vol. 1 (2002): 335–336.

Borchardt, J.K. "The Business of Pharmacogenomics," *Modern Drug Discovery* vol. 4 (2001): 35–39.

Bradsher, Keith. "Vietnam Becomes a Safe Haven for Tourists." *The New York Times*, International section, January 5, 2003.

Buckingham, J.S. "Damascus: A Center of Learning in the Medieval Islamic World," in *Travels Among the Arab Tribes Inhabiting the Countries East of Syria and Palestine* (London: Longman, Hurst, Rees, Orme, Brown and Green, 1825).

Bush, President G.W. "Address to the Nation on Stem Cell Research." Vanderbilt Television News Archives, August 2001.

Buzurg, Ibn Shahriyar, "Men of Science," in *Kitab 'ajayibal-Hind* (Leiden, Netherlands: Brill, 1886).

Cambridge Encyclopedia of the Middle East and North Africa (Cambridge: Cambridge University Press, 1988).

CBS News, April 17, 2002. "New Cure for Bubble Boy Disease" (www.cbs.com).

CDC Fact Sheets on Biologic and Nuclear Terrorism, as of November 15, 2002 (www.cdc.gov).

Cervantes, R.B., J.R. Stringer, C. Shao, J.A. Tischfield and P.J. Stambrook. "Embryonic Stem Cells and Somatic Cells Differ in Mutation Frequency and Type." *Proceedings of the National Academy of Sciences, USA* 99 (2002): 3586–3590.

Chapman, A.R., M.S. Frankel and M.S. Garfinkel. "Stem Cell Research and Applications: Monitoring the Frontiers of Biomedical Research." *AAAS/ICS Report*, 1999.

Chemical Casualty Care Division, *Medical Management of Chemical Casualties Handbook*, third edition (USAMRICD, Aberdeen Proving Ground, Maryland, July 2000).

Davis, S. and Bill Davidson. *2020 Vision: Transform Your Business Today to Succeed in Tomorrow's Economy* (New York, NY: Fireside, 1991).

Dawkins, Richard. *A River Out of Eden* (London: Basic Books, reprint edition, 1996).

Diamond, Jared. *Germs, Guns, and Steel: The Fates of Human Societies* (New York, NY: W.W. Norton 1997).

Discovery Channel. "Raising the Mammoth" 2000 (www.discovery.com).

Dowd, Maureen. "From Botox to Botulism." *New York Times*, Opinion-Editorial (September 25, 2001).

Drews, J. "Drug Discovery: A Historical Perspective." *Science* vol. 287 (2000): 1960–1964.

Enriquez, Juan. "Too Many Flags." *Foreign Policy* (Fall, 1999).

Enriquez, Juan and Ray Goldberg. "Transforming Life, Transforming Business: The Life Science Revolution." *Harvard Business Review* (April/May 2000).

Enriquez, Juan and Rodrigo Martinez. *The Biotechonomy (1.0): A Rough Map of Biodata Information Flow*. Harvard Business School Working Paper (Winter 2002).

Evans, M.J. and M.H. Kaufman. "Establishment in Culture of Pluripotential Cells from Mouse Embryos." *Nature* 292 (1981): 154–156.

Fahnestock, S.R. "Fibrous Proteins from Recombinant Microorganisms," in *Biopolymers* vol. 8, S. Fahnestock and A. Steinbuches (eds), *Polyamides and Complex Proteinaceous Materials* (Weinheim, Germany: Wiley-VCH, 2003).

Farley, M. "Roman Catholic Views on Research Involving Human Embryonic Stem Cells," in S. Holland, K. Lebacqz and L. Zoloth (eds), *The Human Embryonic Stem Cell Debate* (Cambridge, MA: MIT Press, 2001), 113–119.

Fields, S. "The Interplay of Biology and Technology." *Proceedings of the National Academy of Sciences* vol. 98 (2001): 10051–10054.

Forrer, P., S. Jung and A. Pluckthun. "Beyond Binding: Using Phage Display to Select for Structure, Folding and Enzymatic Activity in Proteins." *Current Opinion in Structural Biology* vol. 9, no. 4 (1999): 514–20.

Friend, T. "Half of Stem Cell Money Could Go to Royalties." *USA Today* August 13, 2001, A1.

Fritz, M., A.M. Belcher, M. Radmacher, D.A. Walters, P.K. Hansma, G.D. Stucky, D.E. Morse and S. Mann. "Flat Pearl from Biofabrication of Organized Composites on Inorganic Substrates." *Nature* vol. 371 (1994): 49–51.

Fritz, M. and D.E. Morse. "The Formation of Highly Organized Biogenic Polymer/Ceramic Composite Materials: The High-performance Microlaminate of Molluscan Nacre." *Current Opinion in Colloid and Interface Science* vol. 3, no. 1 (1998): 55–62.

Gearhart, J. "New Potential for Human Embryonic Stem Cells." *Science* 282 (1998): 1061–1062.

Grody, W. "Cystic Fibrosis: Molecular Diagnosis, Population Screening and Public Policy." *Archives of Pathology and Laboratory Medicine* 123 (1999).

Haberman, A.B. *Biochemical Pathway and System Analysis for Target Identification and Validation.* CHI Reports, Cambridge Healthtech Institute, April 24, 2002.

Haberman, A.B., G.R. Lenz and D.E. Vaccaro. *From Data to Drugs: Strategies for Benefiting from New Drug Discovery Technologies.* Scrip Reports, Morpace Pharma Group and PJB Publications Ltd, July 1999.

Haberman, A.B., D. Lockwood and M.A. Branca. *Target Identification and Validation: Key Approaches for Improving the Efficiency and Profitability of Drug Discovery and Development.* CHI Reports, Cambridge Healthtech Institute, June 30, 2001.

Hal, Varian and Peter Lyman. *How Much Information?* December 2000 (http://www.sims.berkeley.edu/research/projects/how-much-info/).

Harvard University and the Nuclear Threat Initiative. *Securing Nuclear Weapons and Materials: Seven Steps for Immediate Action* (Washington, DC, 2002).

Hayflick, Leonard. *How and Why We Age* (New York, NY: Ballantine Books, 1994).

Hodgson, Marshall G.S. *The Venture of Islam*, vol. 2 (Chicago, IL: University of Chicago Press, 1974).

Huxley, Aldous. *Brave New World* (Garden City, NY: Doubleday, Doran and Company, Inc., 1932).

Institute for Genomic Research. "Scientists Decipher the Genetic Code of Malaria." Rockville, Maryland, October 2, 2002 (TIGR.org).

International Petroleum Encyclopedia, 2002. Basic Petroleum Data Book, Petroleum Industry Statistics, vol. XXII, no. 2, August 2002.

International Service for the Acquisition of Agribiotech Applications (ISAAA) Report, January 14, 2003.

Islamic Medical Manuscripts at the National Library of Medicine, Washington, DC. (http://www.nlm.nih.gov/hmd/arabic/arabichome.html).

Kaitin, K.I. "Pharmaceutical Industry Innovation is More Dispersed Despite M&A Activity." *Tufts Center for the Study of Drug Development Impact Report* vol. 2, no. 5 (June 2000).

Kaitin, K.I. and E.M. Healy. "The New Drug Approvals of 1996, 1997 and 1998: Drug Development Trends in the User Fee Era." *Drug Information Journal* vol. 34 (2000): 1–14.

Kaku, Michio. *Visions: How Science will Revolutionize the 21st Century* (New York, NY: Anchor Books, 1997).

Kim, Jong-Hoon et al. "Dopamine Neurons Derived from Embryonic Stem Cells Function in an Animal Model of Parkinson's Disease." *Nature* 418 (2002): 50–56.

Kirk, O., T.V. Borchert and C.C. Fuglsang. "Industrial Enzyme Applications." *Current Opinion in Biotechnology* vol. 13, no. 4 (2002): 345–351.

Kolata, G. "Pushing Limits of the Human Lifespan." *New York Times*, March 9, 1999, page D1.

Layne, Scott P., Tony J. Beugelsdijk and C. Kumar N. Patel (eds). *Firepower in the Lab: Automation in the Fight Against Infectious Diseases and Bioterrorism* (Washington, DC: Joseph Henry Press, 2001).

Lederberg, Joshua (ed.). *Biological Weapons: Limiting the Threat* (Belfer Center for Science and International Affairs, Harvard University, MIT Press, 1999).

Lewis, Bernard. *What Went Wrong? Western Impact and Middle Eastern Response* (Oxford: Oxford University Press, 2002).

Lumelsky, Nadya et al. "Differentiation of Embryonic Stem Cells to Insulin-Secreting Structures Similar to Pancreatic Islets." *Science* 292 (2001): 1389–1394.

Macer, D. et al. "International Perceptions and Approval of Gene Therapy." *Human Gene Therapy* 6 (1995).

Maddison, Angus. *Monitoring the World Economy 1820–1992* (Paris: OECD, 2000).

Maddison, Angus. *The World Economy* (Paris: OECD, 2001).

Manne, S., C.M. Zaremba, R. Giles, L. Huggins, D.A. Walters, A. Belcher, D.E. Morse, G.D. Stucky, J.M. Didymus, S. Mann and P.K. Hansma. "Atomic Force Microscopy of the Nacreous Layer in Mollusc Shells." *Proceedings of the Royal Society of London* vol. 256, no. 1345 (1994): 17–23.

Marshak, D.R., R.L. Gardner and D. Gottleib (eds). *Stem Cell Biology*. Cold Spring Harbor Monograph Series 40 (New York, NY: Cold Spring Harbor Laboratory Press, 2000).

Martin, G.R. "Isolation of a Pluripotent Cell Line From Early Mouse Embryos Cultured in Media Conditioned by Teratocarcinoma Stem

Cells." *Proceedings of the National Academy of Sciences, USA* 78 (1981): 7634–7638.

Matsui, Y., K. Zsebo and B.L. Hogan. "Derivation of Pluripotential Embryonic Stem Cells from Murine Primordial Germ Cells in Culture." *Cell* 70 (1992): 841–847.

McDonald, J.W. et al. "Transplanted Embryonic Stem Cells Survive, Differentiate, and Promote Recovery in Injured Rat Spinal Cord." *Nature Medicine* 5 (1999): 1410–1412.

"Medical Images," National Library of Medicine collection, Washington, DC.

Michaels, A. "Pfizer R&D Unable to Sustain Group Growth Rate." *Financial Times*, September 12, 2001, 30.

Military Medical Operations Office, *Medical Management of Radiological Casualties Handbook* (Armed Forces Radiobiology Research Institute, Bethesda, Maryland, December 1999).

MIT News. "Technique Could Improve Cartilage Repair," July 18, 2002 (Thomson@mit.edu).

Nasr, Seyyed Hossein. *Islamic Science: An Illustrated Study* (World of Islam Festival Publishing Co., 1976).

National Bioethics Advisory Commission. *Ethical Issues in Human Stem Cell Research* (US Government Printing Press, 1999).

National Renewable Energy Laboratory (NREL). "Bioethanol – Moving into the Marketplace," in *Biofuels for Sustainable Transportation* (Washington, DC: NREL. 2001).

National Research Council (NRC). "Processing Technologies," in *Bio-based Industrial Products – Priorities for Research and Commercialization* (Washington, D.C: National Academy Press, 2000), 74–102.

National Research Council (NRC). "Range of Bio-based Products," in *Bio-based Industrial Products – Priorities for Research and Commercialization* (Washington, DC: National Academy Press, 2000), 55–73.

Nature Genetics, January 1999.

"Nature Insight: Stem Cells." *Nature* vol. 414, no. 6859 (November 1, 2001): 88–132.

Nature magazine, October 2002.

Negroponte, Nicholas. *Being Digital* (New York, NY: Vintage Books, 1996).

Nestle, Marion. "Allergies to Transgenic Foods: Questions of Policy." *New England Journal of Medicine*, March 14, 1996.

New England Journal of Medicine, April 2002.

New York Times, December 7, 1993.

New York Times. "Whether to Make Perfect Humans." July 22, 1982.

Nordlee, Julie A., Steve L. Taylor, Jeffrey A. Townsend, Laurie A. Thomas and Robert K. Bush. "Identification of a Brazil-Nut Allergen in Transgenic Soybeans." *New England Journal of Medicine*, March 14, 1996.

Okarma, T.B. "Human Embryonic Stem Cells: A Primer on the Technology and its Medical Applications," in S. Holland, K. Lebacqz and L. Zoloth (eds) *The Human Embryonic Stem Cell Debate* (Cambridge, MA: MIT Press, 2001), 3–14.

Oliver, Richard W. *The Coming Biotech Age: The Business of Bio-Materials* (New York, NY: McGraw-Hill, 2000).

Olson, Steve. *Mapping Human History: Discovering Our Past Through Our Genes* (New York, NY: Houghton Mifflin, 2002).

Organisation for Economic Co-operation and Development (OECD). "Enzymatic Production of Acrylamide (Mitsubishi Rayon, Japan)," in *The Application of Biotechnology to Industrial Sustainability – Sustainable Development* (Paris: OECD, 2001), 71–76.

Organisation for Economic Co-operation and Development (OECD). "Manufacture of Riboflavin (Vitamin B2) (Hoffmann, La-Roche, Germany)," in *The Application of Biotechnology to Industrial Sustainability – Sustainable Development* (Paris, France: OECD, 2001), 51–55.

Organisation for Economic Co-operation and Development (OECD). "Industrial Sustainability and the Role of Biotechnology," in *Biotechnology for Clean Industrial Products and Processes – Towards Industrial Sustainability* (Paris: OECD, 1998), 15–27.

Organisation for Economic Co-operation and Development (OECD). "Polymers from Renewable Resources (Cargill Dow, United States)," in *The Application of Biotechnology* (2001), 87–90.

Oxford Encyclopedia of the Modern Islamic World (New York, NY: Oxford University Press, 1995).

Resnick, J.L., L.S. Bixler, L. Cheng and P.J. Donovan. "Long-term Proliferation of Mouse Primordial Germ Cells in Culture." *Nature* 359 (1992): 550–551.

Rideout III, William M. et al. "Correction of a Genetic Defect by Nuclear Transplantation and Combined Cell and Gene Therapy." *Cell* 109 (2002): 17–27.

Rifkin, Jeremy. *The Biotech Century: The Coming Age of Genetic Commerce* (London: Victor Gollancz, 1998, and New York, NY: Jeremy Tarcher, Inc., 1998).

Roberts, R.W. and W.W. Ja. "In Vitro Selection of Nucleic Acids and Proteins: What are we Learning?" *Current Opinion in Structural Biology* vol. 9, no. 4 (1999): 521–529.

Schrödinger, Erwin. *What is Life? The Physical Aspect of the Living Cell* (Cambridge: Cambridge University Press, 1948).

Science, 279 (1998).

Scientific American, September 1996.

The Scientist, vol. 14, no. 21, October 30, 2000.

Scott, Bruce R. and George C. Lodge (eds). *US Competitiveness in the World Economy* (Boston: Harvard Business School Press, 1985).
Shamblott, M.J. et al. "Derivation of Pluripotent Stem Cells from Cultured Human Primordial Germ Cells." *Proceedings of the National Academy of Sciences, USA* 95 (1998): 13726–13731.
Shimizu, K., J. Cha, G.D. Stucky and D.E. Morse. "Silicatein α: Cathepsin L-like Protein in Sponge Biosilica." *Proceedings of the National Academy of Sciences*, Washington, DC, vol. 95, no. 11 (1998): 6234–6238.
Sifton, D.W. (ed.). *PDR Guide to Biological and Chemical Warfare Response* (Thompson/Physicians' Desk Reference, 2002).
Solbrig, Otto, Robert Paarlberg and Francesco Di Castri. *Globalization and the Rural Environment* (Cambridge, MA: Harvard University Press, 2001).
Sorio, B. et al. "Insulin-Secreting Cells Derived from Embryonic Stem Cells Normalize Glycemia in Streptozotocin-induced Diabetic Mice." *Diabetes* 49 (2001): 157–162.
Spemann, H. (1943). "Forschung und Leben" (Research and Life). Quoted in Horder, T.J., Witkowski J.A. and Wylie, C.C. *A History of Embryology* (Cambridge: Cambridge University Press, 1986), 219.
Steen, G. *DNA and Destiny: Nature and Nurture in Human Behavior* (New York, NY: Plenum, 1996).
Stock, G. *Redesigning Humans: Our Inevitable Genetic Future* (Boston, MA: Houghton Mifflin, 2002).
"The Storefront Genome." Conference organized by the Center for Society, the Individual and Genetics (http://www.arc2.ucla.edu/csig/symp1.htm).
Sykes, Bryan. *Seven Daughters of Eve* (New York, NY: W.W. Norton, 2001).
Teresi, Dick. *Lost Discoveries: Ancient Roots of Modern Science from the Babylonians to the Maya* (New York, NY: Simon & Schuster, 2002).
Testimony of Dr Sue Bailey before Congress: The United States House of Representatives Committee on Veterans' Affairs, October 15, 2001.
Thomson, J.A. et al. "Embryonic Stem Cell Lines Derived from Human Blastocysts." *Science* 282 (1998): 1145–1147.
Tufts Center for the Study of Drug Development, press release, November 30, 2001 (http://csdd.tufts.edu/).
University of Texas Southwestern Medical Center. "Immortalizing Enzyme Does Not Make Human Cells Cancerous," January 4, 1999.
US Army Medical Research Institute of Infectious Diseases. *Medical Management of Biological Casualties Handbook*, third edition (Fort Detrick, Maryland, July 1998).
US Army Medical Research and Matériel Command. *Defense Against Toxin Weapons*, second revision (Fort Detrick, Maryland, 1997).
Viney, C., A.E. Huber, D.L. Dunaway, K. Kerkam and S.T. Case. "Optical Characterization of Silk Secretions and Fibers," in D. Kaplan, W.W. Adama,

B. Farmer and C. Viney (eds), "Silk Polymers," *ACS Symposium Series*, vol. 544, American Chemical Society, Washington, DC (1994), 120–136.

Vollrath, F. and D.P. Knight. "Liquid Crystalline Spinning of Spider Silk." *Nature* vol. 410 (2001): 541–548.

Warner, S. "Science in Small Hubs." *The Scientist* vol. 16, no. 7 (April 1, 2002): 54.

Watson, James and Francis Crick. "A Structure for Deoxyribose Nucleic Acid." *Nature* vol. 171, April 25, 1953.

Weber, M.F., C.A. Stover, L.R. Gilbert, T.J. Nevitt and A.J. Ouderkirk. "Giant Birefringent Optics in Multilayer Polymer Mirrors." *Science* vol. 287 (2001): 2451–2456.

Whaley, S.R., D.S. English, E.L. Hu, P.F. Barbara and A.M. Belcher. "Selection of Peptides with Semiconductor Binding Specificity for Directed Nanocrystal Assembly." *Nature* vol. 405, no. 6787 (2000): 665–668.

Whitehead Institute for Biomedical Research. "New Study Shows Normal-Looking Clones May Be Abnormal," MIT, Cambridge, MA, July 9, 2001 (www.wi.mit.edu).

Wiegele, Thomas C. *Biotechnology and International Relations: The Political Dimensions* (Gainesville, FL: University of Florida Press, 1991).

Willcox, P.J., S.P. Gido, W. Muller and D.L. Kaplan. "Evidence of a Cholesteric Liquid Crystalline Phase in Natural Silk Spinning Processes." *Macromolecule*, vol. 29 (1996): 5106–5110.

Wolpe, P.R. and G. McGee. " 'Expert Bioethics' as Professional Discourse: The Case of Stem Cells," in Holland, S., Lebacqz, K. and Zoloth, L. (eds), *The Human Embryonic Stem Cell Debate* (Cambridge, MA: MIT Press, 2001), 204–242.

World Bank. *World Development Indicators 2002* (CD ROM).

Zaremba, C.M., A.M. Belcher, M. Fritz, Y.L. Li, S. Mann, P.K. Hansma, D.E. Morse, J.S. Speck and G.D. Stucky. "Critical Transitions in the Biofabrication of Abalone Shells and Flat Pearls. *Chemistry of Materials* vol. 8 (1996): 679–690.

Zimmerman, Burke E. "Human Germ-Line Therapy: The Case for its Development and Use." *Journal of Medicine and Philosophy*, December 1991.

INDEX